Pluto

My Journey from Last to First

Kevin M. Caruso

Pluto: My Journey from Last to First
ISBN# 978-1-7372694-8-9 (Science, Education, Non-Fiction, Paperback)
© Copyright 2023 by Kevin M. Caruso

All rights reserved. Proudly Printed in the United States of America. No part of this publication may be reproduced, stored in a retrieval system, or translated in any form or by any means, electronic, mechanical, photocopying, recording, scanning, or otherwise except as permitted under United States Copyright Law, without the prior written permission of the author and copyright owner.
The Author asserts no copyright for non-author photos, drawings, "text in quotes", news article p.17.

Acknowledgments: The author is grateful to the individuals and organizations credited throughout this work for use of their photographs, particularly: Federal Aviation Administration, Hubble Space Telescope Science Institute, Jet Propulsion Laboratory/Caltech, Johns Hopkins University Applied Physics Laboratory, Lowell Observatory, NASA, Smithsonian Institution, Southwest Research Institute, Subaru Telescope, U.S. Naval Observatory, U.S. Postal Service, Wikipedia. Thank you all!
(Book Front Cover Image & Inside Cover Page Credits: NASA/JHUAPL/SwRI/ZLDOYLE). 5-823r

Inquiries should be addressed/emailed to:

Kevin M. Caruso, Author & Publisher
502 South Pine Street
Mount Prospect, Illinois 60056, U.S.A.
kcaruso1994@gmail.com

This book is dedicated with love to:

God, for a lifetime of blessings and learning
Terese
Mary & Andy, Maggie, Karma and the entire Sullivan family
Steven & Heather w/pups Bentley, Dakota, Skye
Kristina & Jeremy with Hayden, Breanna, Dylan, Jescie & Cameron
Colleen
Tim and Gina with Lia & Gianna
Charlotte
Alex
Thank you for your love throughout this journey on planet Earth!

Notes: 1. The Table of Contents includes the word "AND" after each Chapter Subject. There was a funny television commercial years ago which recommended using the word "AND" after all requests, effectively ASKING for MORE, a bonus request. The concept is used in reverse here: Instead of asking for more, You'll be Receiving More: a Bonus Subject in addition to the main chapter content, Plus Pluto's Practical Lessons.
2. **Bold and Underlined Words** are defined in the Giant Glossary/Index
3. Chapter and Lesson Numbering fall "outside the box"…like Pluto!

Table of Contents

Copyright Page, Dedication, Explanation of this Table of Contents (page 1 bottom) 1

Introduction: Pluto: My Journey from Last to First 3

Chapter 1: Clyde Tombaugh, Qualifying Himself for Astronomy (AND Wright Brothers' Approach) 4

Chapter 2: Percival Lowell, Laying the Foundation (AND Ronstadt's Corridor Principle) 8

Chapter 3: The Blink of an Eye! (AND Dr. Carl Sagan's "Pale Blue Dot") 12

Chapter 3.5: Venetia Burney, Naming a Planet for Breakfast (AND Pluto's Fun Fractional Birthdays) .. 18

Chapter 3.7: You Just Discovered Pluto, Now What? (AND NASA's Mars Helicopter Ingenuity) 22

Chapter 4: James Christy, Honoring Mrs. Christy (AND Cousteau's Life/Aqua-Lung) 26

Chapter 5: Hubble Space Telescope, 5 Moons! (AND Continuous Improvement) 29

Chapter 6: Alan Stern, 29-Cent Challenge...Pluto Explored! (AND Rohn's Ant Philosophy) 32

Chapter 7: IAU Astronomers, Defining "Planet" (AND the Law of Cause and Effect) 36

Chapter 7.5: Science is about Disagreement! (AND Edison's 10,000 Failures) 41

Chapter 8: A Heart Bigger Than Texas! (AND Improved Pluto Postage at Last!) 44

Chapter 9: New Horizons, Rocking the Solar System (AND Einstein's Exam Fable) 49

Chapter 10: Honoring Pioneers! (AND Disney's EPCOT & Carnegie/Hill's Philosophy of Success) 56

Chapter 11: Next Goal? Arrokoth and Beyond! (AND Tracy/Ziglar's Goal Setting) 60

Chapter 12: Spacecraft Science (AND Two Special Shapes in the World) 62

Chapter 13: Success and Bravo! (AND What did New Horizons Pack for Pluto?) 65

Chapter 14: Lessons Learned, Kuiper Belt Wisdom (AND Relevant Quotes) 66

Giant Glossary/Index: Bold and underlined words are defined and located by page 80

Dual Bibliography: Clyde, Pluto, Astronomy (AND Success!) 93

Awesome Web Sites for Ongoing Education re: Clyde, Pluto, Astronomy (AND Successful Living) 95

Boldly Go! (a beneficial philosophy for both Science Fiction and Science Fact!) 97

Author Biography and Gratitude 98

Introduction: Pluto: My Journey from Last to First

After 76 years as the 9th planet (from discovery February 18, 1930 to reclassification August 24, 2006), astronomers of the **International Astronomical Union (IAU)** vigorously debated and quickly voted on their first-ever scientific definition of the word "planet". **Pluto was re-classified from planet to newly-coined "dwarf planet"---a re-classification based upon a wealth of mounting observable evidence.**

For their 17th annual word of the year vote, the **American Dialect Society** humorously chose the word **"Pluto-ed"** as their 2006 Word of the Year. Getting **"Pluto-ed" or "Plutoed"** meant:
1. To be demoted without due cause or reason.
2. To demote or devalue someone or something with considerable cause.
3. To demote or devalue someone or something, as happened to the former planet Pluto when the General Assembly of the IAU decided Pluto no longer met its definition of a planet.

SCIENCE however is neither Opinion, Demotion nor Promotion. It's data-driven: observing nature and offering scientific theories which best fit observations...until those theories are obsoleted by new ones which better fit the data. **Science is a transparent open dynamic process: The Scientific Method (p 42).**

Years after Pluto's 1930 Discovery, astronomers offered new theories suggesting the existence of more objects---similar to, but beyond---Pluto, because Pluto is different than the other planets:
- **Pluto isn't rocky like the inner planets: Mercury, Venus, Earth and Mars.**
- **Pluto isn't a gas giant like the outer planets: Jupiter, Saturn, Uranus and Neptune.**
- **Pluto IS a tiny ice world** with a highly inclined oval orbit outside the planetary orbital plane (which brings Pluto closer to the Sun than Neptune for 20 years of its 248-year orbit), with a highly tipped rotational axis, which shares a donut-shaped region of the outer solar system (the Kuiper Belt) with a vast number of similar objects (KBOs), some of which are of comparable size.

First confirmation of Pluto's icy Kuiper Belt neighbors occurred on August 30, 1992 when astronomers David Jewitt and Jane Luu discovered the first Kuiper (rhymes with viper) Belt Object (KBO) 1992QB1, beyond Pluto. They found 2 more in 1993. The Known KBO count grows regularly & is now over 2000!

On October 21, 2003, using the **Palomar Observatory**, astronomers **Michael Brown, Chad Trujillo** and **David Rabinowitz** spotted KBO 2003UB313 (**Xena**, later renamed **Eris**)---**estimated to be larger than Pluto! If Pluto was a planet, then so was Eris,** and so maybe were other large newfound KBOs.

Astronomers had been debating Pluto's "planethood" for years as more KBOs were discovered. On August 24, 2006, at the **26th General Assembly meeting,** the IAU voted to clearly define the word "planet". **The new definition reclassified Pluto (and other objects) as Dwarf Planets.**

The 2006 IAU vote turned out to have a silver lining for Pluto fans! No longer the last or 9th planet in our solar system, Pluto transitioned to Closest or First Kuiper Belt Object in our Solar System---the 1st of the 3rd Great Class of Solar System Objects: (1) inner rocky planets, (2) outer gas giants, (3) Kuiper Belt Objects (KBOs)---**Pluto was now 1st!**

Pluto's Journey is an Ongoing Adventure of Scientific Discovery! Science, Technology, Engineering, Art and Math (STEAM) concepts AND some Kuiper Belt Wisdom are highlighted along the way for you! Pluto's Amazing Journey continues & awaits you in this exciting 2023 snapshot in time!
    Phenomenal Regards! ---Kevin Caruso, Mount Prospect, Illinois Updated May 8, 2023

Chapter 1:
Clyde Tombaugh, Qualifying Himself for Astronomy

"Before everything else, getting ready is the secret of success." – Henry Ford

Clyde Tombaugh discovered **Pluto** on Feb 18, 1930 after 13 months of nightly photographing and reviewing hundreds of thousands of stars from the **Lowell Observatory** in Flagstaff, AZ.

Pictured here is 24-year old Clyde with his 3rd home-built telescope, a 9-inch Newtonian **Reflector** (mirror-based) Telescope which he built in 1928 from scrap farm machinery and a discarded 1910 axle from his father's Buick.

This is the telescope through which he saw Jupiter and Mars and made sketches which he sent to Lowell Observatory, which led to his being hired to search for a 9th planet.

(Credit: Lowell Observatory Archives)

The first of 6 children, Clyde William Tombaugh was born on February 4, 1906 on a farm in Streator, Illinois. Growing up on a farm included helping his parents from sunrise to sunset with the daily work. There he learned the values of hard work, family, honesty, integrity, and gained an interest in geography, history, trigonometry and physics. **Importantly, Clyde developed a great curiosity about the world and the sky, plus a love for reading.**

Clyde's Uncle Lee owned a farm about 9 miles away. He was fascinated with the sky and owned a 3-inch diameter **refractor** (lens-based) **telescope** and some books on **astronomy** which he shared with Clyde's family when they visited.

> **Clyde first looked through his uncle's telescope when he was 12 years old. He was amazed to see mountains and craters on the Moon. This transformed the Moon from a thing in the sky to a place!** From then on, Clyde was "hooked" on astronomy! Remember this at page 65.

> **Lesson 0.5: Engage others in your passion.** Share your passion and your gifts with the world. **Develop an appetite for learning through reading.** Reading daily in your field of interest is the key to becoming highly knowledgeable in that field.

Clyde read more astronomy books and articles to learn as much as possible about the planets. When he was 14 years old, his father and uncle purchased a more powerful refractor telescope, through which Clyde was able to see planet **Jupiter** and the North polar ice cap on **Mars**!

In 1922, for **economic** reasons, the Tombaugh family moved to farmland in Burdett Kansas, also owned by Uncle Lee. Lee gave the bigger telescope to Clyde's family to keep. Clyde continued to read articles about telescope building, and, starting in 1926 at age 20 decided to build his own. **He wrote to other astronomers for advice** regarding telescopes they had built for themselves.

> **Lesson 1: Learn from the experts, the best. Who are the best? Those who have done or who are already doing what you want to do.** Seek them out and ask for their help and advice. **Leverage** their experience in order to shorten your path. Then Take Action in the direction of your goals. Your own action or forward motion is the key to getting results. **And Stay Curious!**

Okay, Time Out. Historical Note:
Clyde sought out experts for their advice and experience just as the **Wright Brothers** did in their approach to studying flight. Wilbur Wright wrote to the **Smithsonian Institution** requesting any available papers they had published about flight. Like Clyde---the brothers had no credentials either when it came to flying. They owned a bicycle shop in Dayton, OH.

In his May 30, 1899 letter to the Smithsonian, Wilbur wrote:

> "I am about to begin a systematic study of the subject in preparation for practical work to which I expect to devote what time I can spare from my regular business...I am an enthusiast, but not a crank in the sense that I have some pet theories as to the proper construction of a flying machine. "

Here's the famous photo of Orville piloting their Flyer as Wilber assists. First Successful Powered Flight on December 17, 1903. (Credit: **Federal Aviation Administration** and **National Air and Space Museum**, Smithsonian Institution, Washington DC)

Wilbur's letter continued: "I wish to avail myself of all that is already known and then if possible add my mite to help on the future worker who will attain final success..."

The brothers studied all they could find about the experiences and the writings of Aviation Pioneers **Otto Lilienthal**, **Samuel Pierpoint Langley**, and **Octave Chanute**.

They encountered challenges such as incorrect published air pressure data, crashes, even an engine manufacturer who refused to sell them an engine for such a "foolish notion" as a flying machine. So, they worked with a local machinist to build their own engine. They also built their own wind tunnel to test their wing and propeller designs.

On December 17, 1903, the brothers made history with their 12-second first powered flight! They completed 4 flights that day, before crashing and damaging their aircraft.

Wilbur and Orville Wright. (Credit: FAA and National Air and Space Museum, Smithsonian Institution, Wash DC) (See https://wright.nasa.gov)

Note: Even after success, they continued to learn, building in 1904 the world's 1st take-off catapult at **Huffman Prairie Field** in Ohio, with a take-off track & 1600-lb weight.

See this AMAZING FILM of the Wright Brothers' 1904 Catapult Take-Off.
Note: Film is MISLABELED 1903 & FIRST FLIGHT---it is Neither Nor 1st plane, but still amazing:
Google: "Wright Brothers Catapult Film Footage"

Back to Clyde...

Clyde built several larger telescopes, patiently grinding his own lenses and mirrors. Uncle Lee was a positive influence in Clyde's life---Lee paid Clyde to build a telescope for him, which he did, and they spotted a comet with it. (Note: Clyde eventually built 30 telescopes over the course of his life.)

> **Lesson 1.5: Share your passion with others,** as Uncle Lee did with Clyde about Astronomy and Telescopes. **You don't know who you may be inspiring by sharing your gifts!** (Mars Prequel p.11)

Using the money from Uncle Lee, Clyde bought materials and used scrap farm machinery and car parts to build his own 400 magnification telescope, which he used to see and sketch Mars and Jupiter.

> **Lesson 2.0: Do what you can with what you have, right where you are.** Innovate. Be creative. Maintain a positive optimistic attitude and move forward. Learn all you can and apply your learning and your talents. Clyde learned all he could about telescope building by reading articles, then built several, earning money for his work. He drew sketches of planets using his large home-build telescope, thereby becoming a serious self-educated amateur astronomer. He enjoyed observing planets. He forged a new identity for himself by stepping forward into astronomy.

Clyde had hoped to attend college in 1928, but a 15-minute afternoon hailstorm destroyed the farm's wheat & oat fields and wiped out the family income & possible tuition money for the foreseeable future. Greatly disappointed, Clyde kept moving forward with a positive attitude. He was frustrated with farming because the weather played such a big role in determining the success or failure of the farm.

He continued helping his family earn money by working hard for neighboring farmers whose crops were not destroyed by the storm. (Note: When you reach Chapter 4, compare Clyde's storm adversity with Cousteau's accident adversity & compare their attitudes & decisions to maintain forward momentum).

Lesson 2.5: Expect Obstacles and Adversity on the path toward your goals. Choose and maintain a positive optimistic attitude daily.
Keep moving toward your goals. (right photo credit: <u>Kevin Caruso</u>)

Reinterpret or reframe your negative circumstances and obstacles---what are they trying to teach you? Look for and glean their lessons.

Remember This: Everything Counts! The good, the bad, the obstacles, and progress. Keep preparing yourself, improving, learning from every experience. Keep moving steadily forward.

"In the middle of every difficulty lies opportunity." – <u>Albert Einstein</u>

"A smooth sea never made a skillful mariner." – English Proverb

"We may encounter many defeats but we must not be defeated." - <u>Maya Angelou</u>

"You have within you right now, everything you need to deal with whatever the world can throw at you." – <u>Brian Tracy</u>

"Never give up! Never, Never Give Up!" – <u>Sir Winston Churchill</u>

Clyde remained fascinated with the sky and <u>planets</u> and spent hours watching and making sketches of Jupiter and Mars using his home-built telescope. **He wanted to become an astronomer, but knew he lacked the formal education needed for the job. So he decided to pursue a path as an astronomical observer---here he had plenty of experience.** He had built several good telescopes which permitted him to see the Moon, Mars, and Jupiter. **He read Mars observation reports written by the Lowell Observatory Staff in the magazine "Popular Astronomy", so Clyde mailed some of his sketches to the Lowell Observatory in Flagstaff, Arizona for advice about observing. Remember this: he took action!**

Lesson 2.7: Taking action in the direction of your goals is like initiating the automated <u>launch sequence</u> of accomplishment.

Lesson 3: Be Authentic, Resourceful and Creative (use the tools and skills you have right where you are) and continue to learn. Clyde was a hard-working farm boy who read about telescopes, wrote to a telescope builder, built his own telescopes from old farm machinery and car parts. Then he drew sketches of Jupiter and Mars from his observations. He continued to read and learn about astronomy. He sent his sketches to the astronomers at Lowell Observatory for their feedback.
His personal efforts led to a job offer, to his becoming the discoverer of a new world, to a university scholarship, to meeting his future wife, to a second career teaching in the military, and to a life unimagined. Be open to such possibilities! Anything is possible!

Chapter 2: Percival Lowell, Laying the Foundation

"Luck is what happens when preparation meets opportunity." – Elmer Letterman

Percival Lowell observing Venus during the day, Oct.17, 1914, using his 24" Clark Refractor. He initiated 3 searches for Planet "X" (Credit: Lowell Observatory Archives)

Lowell Observatory's 13-Inch Dome in Flagstaff, Arizona, where Clyde Tombaugh photographed Pluto, leading to discovery. (Credit: The Author, 2021 photo)

Unknown to Clyde, the astronomers at Lowell Observatory had decided to continue the work of their founder Percival Lowell to search for a 9th planet. They were seeking to hire someone to operate the telescope and take long exposure photos of the night sky during the cold winter nights in Flagstaff.

The Lowell astronomers **were so impressed with Clyde's drawings, initiative, and enthusiasm for astronomy, that they asked him about his health and willingness to work long cold nights for low pay in a cold dome, making good observations.** This sounded like a great opportunity for Clyde to pursue his passion for astronomy and observing. After all, he was a newly self-taught amateur astronomer and was familiar with building telescopes and making observations of planets. This was a perfect fit!

They invited Clyde to work at the Observatory for $125 per month. Vesto "V.M." Slipher, Astronomer & Director of the Lowell Observatory after Percival Lowell's death, described Clyde as "a young man from Kansas with enthusiasm and the ability to spend long hours in a cold dome completing good observations." Slipher served as director at Lowell from 1916 to 1954. **His work in spectroscopy increased understanding of enormous galactic velocities & the concept of universal expansion.**

> **Lesson 4: Be Willing to Act Without Formal Credentials!** Clyde went to work at a state-of-the-art Observatory based upon nothing more than a passion for astronomy, books he read, his self-taught skills as a telescope builder, and some sketches of the planets he observed. Clyde eventually created his own credentials by having faith to take action in the direction of his goals. **Act, move forward to see doors of Opportunity. Boldly Go! Step forward and take action to become the person you want to become!** Clyde had activated the Corridor Principle, without knowing it by that name.

Corridor in the lower level of the Library of Congress, Washington DC which illustrates the Corridor Principle. (Credit: Kevin Caruso)

Until you start walking down the corridor or path of life which you alone choose, you cannot not even see the doors of opportunity which await you on either side.

You alone choose whether to start down this corridor. Your action/motion down the hall starts the seeing process.

The Corridor Principle was first defined by <u>Dr. Robert Ronstadt,</u> professor of Babson College's <u>Entrepreneurship</u> program in the late 1970's. He published an article on 8-1-2007 in the publication "Viewpoint".

Lesson 4.5: Get familiar with the Corridor Principle. Dr. Robert Ronstadt of Babson College's Entrepreneurship Program in the late 1970s developed and taught a concept called the "**Corridor Principle**". In an article he published, he described the Wright brothers, McDonald brothers, Bill Gates and other entrepreneurs as having something in common. They started earlier ventures and leveraged (there's that word again) and applied their prior experiences to new opportunities that arose toward their new venture. The Corridor Principle requires that a person take action---like walking down a corridor or hallway or path (toward a goal which we choose).

Hidden Possibilities and "Doors" of Opportunity only become visible as we start moving along a "corridor" or path toward a goal which we choose, while maintaining a positive, seeking, grateful, expectant attitude. KEY: Doors often lead to Unanticipated Successes in areas not originally pursued!

If you stay at the beginning of the corridor, you can't see the many doors of opportunity on either side of the hall. But once you start walking down the corridor, these possibilities and doors of opportunity will become visible to you---they were there at all times, but you'd never have seen them had you not started walking down your chosen path (or corridor). You must be in motion!

And a positive attitude is essential. In his audio learning program entitled "**The Essence of Success**", author and co-founder of the **Nightingale-Conant Corporation** of Chicago (www.nightingale.com) **Earl Nightingale** said **"Attitude"** is the magic word. He said that the world reflects back to us the attitude that we project to the world. And he further stated that attitude is a choice we make each day. It's a rephrasing of the biblical principal: What you send out, you get back. Or what you sow, you'll reap. In flying terms, a proper wing attitude maintains <u>lift</u> and keeps a plane flying. So how's your attitude?

Let's back up 35 years to 1894. Percival Lowell (1855-1916) was a wealthy Boston businessman **and a self-taught astronomer.** That year, he built a private observatory in Flagstaff, Arizona to study planet Mars, starting with a borrowed 18" diameter telescope through which he saw what appeared to be canals on Mars. **He drew detailed sketches of his Mars' observations (see sketches on a Mars globe).**

In 1877, an Italian Astronomer **Giovanni Virginio Schiaparelli** (1835-1910) began mapping Mars' dark and lighter colorations from the **Brera Observatory** in Milan, Italy. He described **"canali"** an Italian word for channels.

Lowell was fascinated with Mars and observed it frequently from 1894 to 1908. He **theorized** that intelligent Martian civilizations had built canals to bring melting water from the Martian polar ice caps to drier equatorial regions. **Lowell's sketches (see right)** from his Mars observations showed canals which appeared to end in dark areas, which he claimed might be vegetation.

In 1895 Lowell published his 1st of 3 books on Mars which promoted his observations & theories:
- *Mars* (published in 1895)
- *Mars and Its Canals* (published in 1906)
- *Mars as the Abode of Life* (published in 1908).

Lowell was **ridiculed** for his theories about Martian Civilization-built canals.

Credit: American Museum Journal

Despite the ridicule, Lowell was an excellent astronomer! Through his careful telescopic observations of Neptune, he believed that Neptune's orbit was being gravitationally altered by some unknown outlying mass---possibly a 9th planet. The thought of discovering a new world beyond Neptune began to **consume** Lowell's thoughts. Discovering a 9th planet would bring him great respect, and this became one of Lowell's primary goals. He built a new 24-inch telescope, and in **1905 commissioned a search for a 9th planet, which he referred to as "Planet X".** He conducted a total of 3 searches over 7-years:

1st search from 1905-1907, using 18-inch telescope which required 3 hours of camera exposure)
2nd search from 1909-1912 using the new larger reflector telescope which required just 7 minutes of camera exposure)
3rd search from 1914-1916. **Pluto was photographed in 1915, but no one noticed it!**

> **Lesson 5: Don't permit others to define who you are. That's your job! Ignore the noise and chatter around you.** It doesn't matter what other people say about you. Continue to focus on your goals. People called Pluto small, a planet, a "dwarf-planet", not a real planet. It didn't matter. Even Clyde once said that whatever Pluto is called, it's still there. Well said Sir!

Not finding Planet X greatly disappointed Lowell. **At age 61, Lowell died of a stroke on November 16, 1916. Although Lowell didn't personally find Planet X, he laid a solid foundation for Pluto's discovery.** His widow sued to defund the observatory and litigation lasted for over a decade, reportedly using up most of the money being sought. The search for Planet X was thus halted...for 13 years!

In late 1928, the astronomers at Lowell Observatory decided to begin a new search for Planet X. Clyde's sketches arrived with excellent timing, and **the astronomers decided to hire the "young man with enthusiasm from Kansas": Clyde Tombaugh.**

This job offer was Clyde's ticket toward his passion and away from farm life. He no longer wanted the weather to determine his future.

Clyde took a leap of faith and decided to leave his parents' Kansas farm home. On January 14, 1929 using some of the money he'd earned from helping his neighbor farmers, and with no guarantee of success, he bought a 1-way train ticket to Flagstaff, AZ, and departed Kansas with several sandwiches his mom made for him for the 28-hour train ride, 835 miles from home. If this didn't work out, Clyde didn't have the money for a train trip home. But he took with him the family values instilled in him from the farm: honesty, integrity, willingness to work hard, to work long hours with great attention to detail.

> **Lesson 6: Never Give Up! Take a Leap of Faith, Take Action and Go For It** when a great opportunity presents itself. Pursue your goals. Clyde left everything behind to pursue his passion for astronomy. Even if you have no formal **credentials** (Clyde had none), take action in the direction of your goals, believe in yourself, persist, find a way to pursue your goals.

In his book "The Case for Pluto" author **Alan Boyle** paints a word picture of a self-made young man:

> Clyde Tombaugh...
> - A Country Boy
> - 24 Years Old
> - High School Graduate
> - Hard Worker
> - Couldn't Afford College
> - Went West
> - Willing to work 14-hour winter nights at an observatory
> - Not enough money for return train ride home
> - Discoverer of Planet X (named Pluto)!
>
> That's a Success Story!

Clyde had an excellent philosophy and confidence and belief in himself!

> "Your personal philosophy is the greatest determining factor in how your life works out." – **Jim Rohn**
>
> "Success doesn't come to you... You go to it." – **Marva Collins**
>
> "Just Do It!" – **Nike**
>
> "No! Try Not. Do or Do Not. There is no Try." - **Yoda**

Author's Side Note about Mars: In cooperation with the **National Geographic Society, Brian Grazer and Ron Howard** produced a 3-part television series called *"MARS"*, available on Blu-ray Disc, and in book form: *"Mars: Our Future On The Red Planet"* by Leonard David from National Geographic Books.

Set in the year 2033, the story combines a realistic crewed mission to Mars with **present-day interviews with some of the most brilliant minds in space exploration,** including **Elon Musk, Neil deGrasse Tyson** and NASA astronauts. **The story depicts hazards, heartbreaks, funding concerns, psychological impacts and rewards of the first crewed outpost on Mars. AND included as a bonus: a short video entitled "Before Mars – A Prequel"** which depicts how a neighbor's ham radio hobby & call to an orbiting space station astronaut inspired a young girl to become future Mission Commander, **just as Clyde's Uncle Lee inspired Clyde with astronomy and telescopes.** Check it out when you have a few free evenings!

Chapter 3: The Blink of an Eye!

"Dr. Slipher, I have found your Planet X...I'll show you the evidence."
- Clyde Tombaugh, Discoverer of Pluto

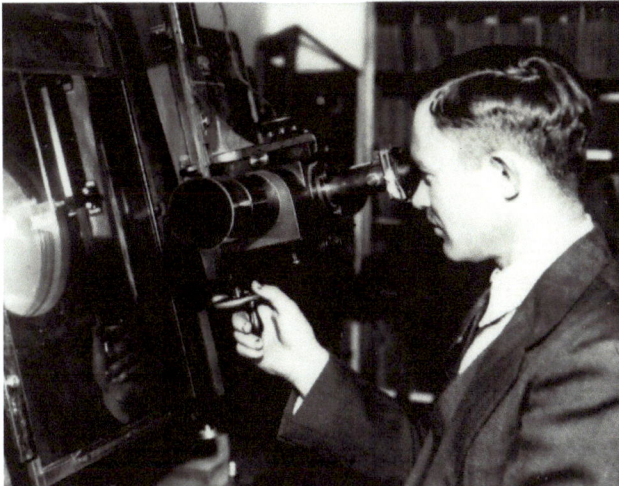 

Clyde peering through Blink Comparator used to discover Pluto.   Blink Comparator on display in Lowell Observatory's Slipher Building Rotunda. (left: Lowell Observatory Archives/Henry Giclas; right: Author)

At Lowell Observatory, Clyde was shown how to photograph the night sky using the telescope and its camera, how to develop the large star field photographic plates of glass, then use a special machine called a **"Blink Comparator"** to compare the star plates. The sky-searching technique involved photographing the same part of the sky on 2 separate nights, several days apart, plus taking an additional 3rd reference image near one of those dates.

Clyde worked in the Lowell Observatory Dome starting in January 1929. The dome housed the telescope and camera equipment. Winter nights in Flagstaff were very cold, and Clyde needed to place logs in the wood-burning stove for heat. At daybreak, he'd add coal to make re-lighting easier that night. His work also included clearing the frequent heavy snows from the observatory dome. **But Clyde was doing something he loved---astronomy, using one of the world's best telescopes and cameras.**

Pluto Telescope. Red Boxing Glove added to prevent head injuries.
Credit: the Author

Lesson 6.5: **Do whatever it takes**---maintaining the fire, clearing the snow---**to get the job done right and in a responsible manner.** People who work in small companies refer to this as "wearing many hats"---hats of various responsibilities, not just one job description.

"What I know is, if you do work that you love, and the work fulfills you, the rest will come." – Oprah Winfrey

"I had always taken seriously one of the adages in one of my school books: 'Do a thing well, or don't do it at all.'" – Clyde Tombaugh

Originally, the senior astronomers took turns "blinking" the photographic plates, but they became busy with many other responsibilities.

That's when Clyde was informed that he'd oversee the entire process, including blinking:

- Photographing the sky
- Developing the photographic glass plates
- Blinking the plates

Clyde recognized this as a huge responsibility of great importance, and he quickly accepted the roles. **He understood very well that the person blinking the plates carried most of the responsibility for finding or not finding the planet.**

> **Lesson 6.9: Prepare yourself in advance** for the increased responsibility which will surely come. And when it does, take that responsibility firmly by the reigns and lead the way.

Clyde photographed an estimated 162,000 stars on every photographic plate in this special way.

- When the Moon was out, the sky would be too bright (this eliminated 2.5 weeks per month).
- When it snowed or rained, he'd have to wait. Weather was still influencing his work.
- When a cloud floated in front of the telescope for his 2nd photograph, the photo set was ruined.
- During down time, Clyde would simply blink previous plates or wait for better conditions.

> **Lesson 7: Work hard. Persevere! Keep moving toward your goals.** This is an example of Hard Work and **PERSEVERANCE**---also the name of the newest Mars Rover (the one which carried the 1st helicopter named **Ingenuity** to Mars).

January 23, 1930 was a night like many others, good for blinking. Clyde focused the telescope and camera on one section of sky and loaded the photographic plate. **Six days later (January 29, 1930) he repeated the process on the same section of sky, so he could later compare the plates.** As a back-up, he also took a 3rd photo of the same section of sky sometime before, between or after the two dates. His 3rd reference photo was taken on January 21st.

> **Lesson 7.2:** "Do not wait; the time will never be 'just right'. Start where you stand, and work with whatever tools you may have at your command, and better tools will be found as you go along." – **Napoleon Hill**

Just before 4:00 PM Mountain Standard Time on February 18, 1930, Clyde loaded the January 23rd and 29th photographic plates into the Blink Comparator and started blinking the images: left, right, left, right, about 3-5 times per second.

DISCOVERY OF THE PLANET PLUTO

January 23, 1930   January 29, 1930

(Credit: Discovery Plates of Pluto, Lowell Observatory Archives)

Pluto was discovered at 4:00 PM local time on February 18, 1930 by 24-year old Clyde William Tombaugh. He saw the faint spec jump from one position to the other (identified by the arrow (added later) in the left and right photos above). Pluto's barely visible in the Jan 29th image!

Because the two pictures showed the same stars in the sky, just a few days apart, **every star in the photo remained still, except for one very faint spec of light.** This appeared to jump from one location to the other---this is how Clyde could identify moving objects like comets, asteroids and planets, against the fixed background of stars.

He was literally shaking with excitement! **"That's It!"**, he said to himself. Because of its **magnitude** (brightness), direction and distance of travel, **Clyde knew this was not an asteroid or comet. It must be a planet---one far beyond Neptune---it must be the Planet X he was seeking!**

> **In Clyde's own words about the discovery of Pluto (credit: "Clyde Tombaugh, Discoverer of the Planet Pluto – Academy of Achievement, 2019"):**
> "It's very **tedious** work and you go through tens of thousands of star images. I came to one place where it actually was. Instantly I knew I had a planet beyond the orbit of Neptune. That was the most instantaneous thrill you can imagine. It just electrified me!"

Clyde needed to be certain, so he looked at a 3rd reference photograph to see if the object was in that photo as well. It was there in exactly the expected place. **In Clyde's words,**

> "I thought I'd better check this third plate, which is another date, see if there's an image there in the right place that would be consistent with the images on the other plates. That was the final proof."

He rechecked additional photos which had been taken. After 45 minutes of verifying his own work and feeling certain of his findings, Clyde notified astronomer **Carl Lampland** whose office was nearby that he had found a Trans-Neptunian planet. Dr. Lampland rushed over to review Clyde's evidence.

Next, Clyde walked to his boss's office and said, "Dr. Slipher, I have found your Planet X." And he continued, "I'll show you the evidence."

> **Lesson 7.5: Express confidence in your work. Trust your results but verify.**

On February 18, 1930, that spec of light was confirmed to be the 9th planet---not yet named. For the next 7 weeks, the Lowell Team tracked, re-photographed and verified the object to confirm its movement as one beyond Neptune.

On March 13, 1930, Percival Lowell's birthday, and the 149th Anniversary of <u>William Herschel</u>'s discovery of the 7th planet Uranus (in 1781), the Lowell Observatory went public with the discovery: A 9th Planet had been discovered! As author <u>Alan Boyle</u> states in his book <u>"The Case for Pluto"</u>, "...in 1930, the telescope, the camera, and the analysis were all giuded by one young man: a country boy who went out West because he didn't want to go into farming and couldn't afford to attend college."

Left: Clyde at the 13-inch telescope eye piece, 1930.
Above: Best photo of Pluto at discovery, 1930.
(Credits: Lowell Observatory Archives)

Here's the best photo of Pluto on the day of discovery---just a spec of light! We've come a long way---see the book cover photo: a false-color fly-by image). The mission's **"fail-safe"** image of Pluto taken by **New Horizons** spacecraft on July 13, 2015, just before closest approach can be found on page 44!

> **Lesson 8: Be patient. Success takes time!** Hard work and Persistence eventually pay off!

> **"Don't you dare underestimate the power of your own instinct."**
> - <u>Barbara Corcoran</u>, Shark on TV's "Shark Tank" show
>
> **"Life begins when you do."**
> - <u>Hugh Downs</u>, American Broadcaster
>
> **Remain a student: We are individually responsible for our education whether we're in school or not.** – Unknown

Let's digress for a few moments about "specs of light" in the sky.

On February 14, 1990, the **Voyager 1** spacecraft passed the orbit of Pluto. Scientists commanded Voyager's camera to turn toward Earth for one last look home. The photo captured Earth itself as a "pale blue dot" in a beam of sunshine, as referred by famous astronomer **Dr. Carl Sagan** in his book **"Pale Blue Dot."** (Credit: NASA) https://solarsystem.nasa.gov/resources/536/voyager-1s-pale-blue-dot/

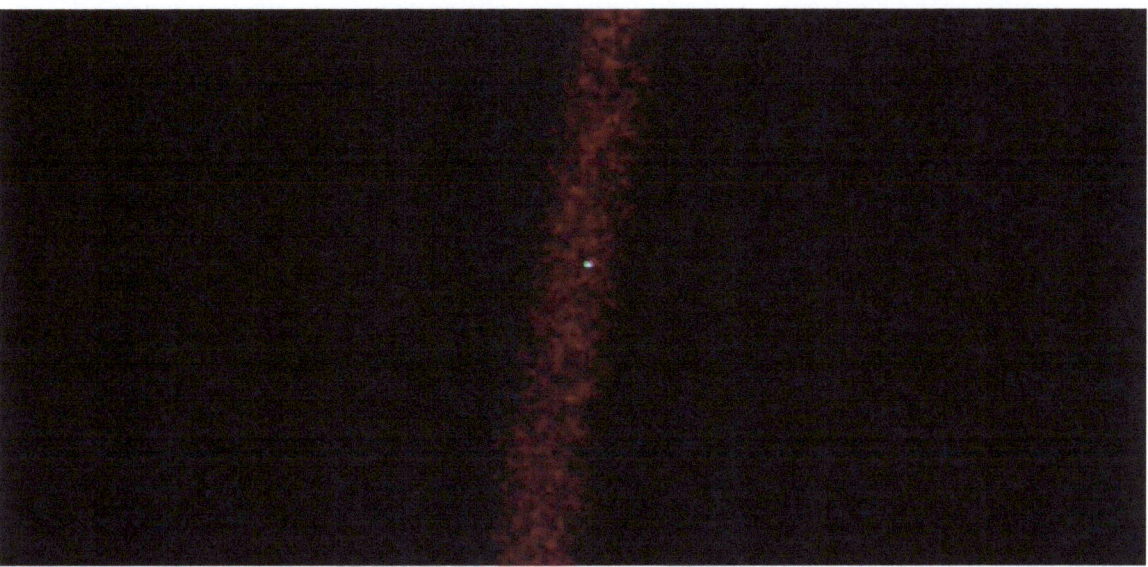

Planet Earth (the "pale blue dot") in a beam of sunshine as described by astronomer Dr. Carl Sagan, as photographed by unmanned spacecraft Voyager 1, as it crossed the orbit of Pluto on February 14, 1990.

While orbiting Saturn on July 19, 2013, the **Cassini Spacecraft** flew into Saturn's shadow. Scientists programmed Cassini to turn around and capture Saturn and Earth in a mosaic photo---another "pale blue dot"---our entire world. (Credit: NASA/JPL-Caltech/SSI) See full Saturn image with multiple planets: https://www.nasa.gov/sites/default/files/pia17172_full_annotated.jpg

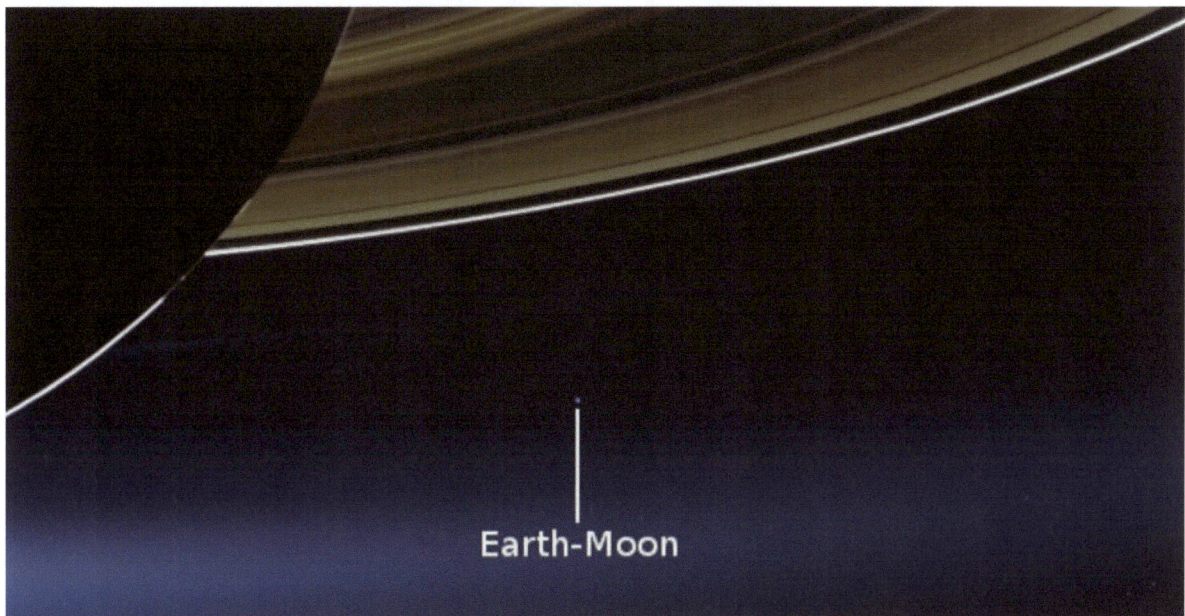

> **Lesson 9: Be Authentic. Remain humble.** Be proud of who you are and where you're from. You are unique in the universe and have an equally unique set of talents and gifts to share with humanity.

And now back to Pluto!

News of discovery of a 9th planet quickly swept the world! Clyde's name was only casually mentioned in the 6th paragraph (see below) of the Associated Press Newspaper Release as a "photographer at the observatory who saw a tiny spot on one of his plates", with paragraph 2 stating that Dr. V.M. Slipher headed the group of "eminent astronomers" who located the "new sphere." **The world quickly hailed Clyde Tombaugh as the discoverer, and he earned recognition as a world-class astronomer!**

FLAGSTAFF, Ariz., March 13 (AP).—In that little cluster of orbs which scampers across the siderial abyss under the name of the polar system there are, be it known, nine instead of a mere eight worlds.

The presence of a ninth marcher in the retinue of the sun, long suspected, was definitely announced here today by Dr. V. M. Slipher of the Lowell observatory, who headed a group of eminent astronomers whose groupings in the milky way with telescopes and cameras located the new sphere.

Size and Age Unknown.

Way out beyond Neptune, tagging bashfully behind his brothers, the new planet's exact whereabouts, size and age are still unknown and it hasn't even got a name. Its presence was mathematically predicted years ago by the late Dr. Percival Lowell, noted scientist who founded the observatory here, partly for the very purpose of identifying it. Other noted astronomers, notably Dr. W. W. Campbell, director of Lick observatory verified Lowell's calculations.

Today the faith in those calculations was rewarded by an announcement by Dr. Slipher that the new planet had been "sighted" last January 21 by an extremely delicate photographic lens, developed for the research. Announcement was withheld, Dr. Slipher said "until we were absolutely sure."

Bigger Than Jupiter.

The discovery revealed that the planet is forty-five times as far from the earth as the earth is from the sun. Although its size has not been definitely determined, it is believed it may be bigger than Jupiter, largest member of the solar family, which is 1,200 times larger than the earth. Astronomers who participated in the discovery are: C. Q. Lampland, E. C. Slipher, J. C. Duncan, K. P. Williams, E. A. Edwards and T. B. Gill.

Until someone entitled to do so gives the sphere a name, it is to be known as "the trans-Neptunian planet." First notice of the body was made by C. W. Tombaugh, photographer at the observatory, who saw a tiny spot on one of his plates. Astronomers soon declared it to be the long sought planet.

(https://www.rarenewspapers.com/view/559832 (left) and /588339 (right))

(Credit: Courtesy of Timothy Hughes, Rare & Early Newspapers, https://www.RareNewspapers.com)

> **Lesson 10:** In your work, don't worry about who gets the credit. Support the Team toward the goal. Truth will reveal itself in time.

> **Lesson 10.5:** Create your own credentials through excellence, hard work and by example.
> Through hard work and persistence and working toward his self-determined goals, **Clyde earned his credentials as a world-class astronomer,** and earned the formal education he couldn't afford when he began his journey to Flagstaff. **Please note that Lowell himself was a self-taught astronomer too.**

Chapter 3.5: Venetia Burney, Naming a Planet for Breakfast

"Why not call it Pluto?" – Venetia Burney at age 11

In his book "**The Pluto Files**", astronomer-author **Neil deGrasse Tyson** states that on March 14, 1930, the day after the 9th Planet was announced by the Lowell Observatory, the name "Pluto" was suggested over breakfast in Oxford England by an 11-year old schoolgirl named **Venetia Burney**.

Venetia Burney, age 11
(Credit: Wikipedia)

Venetia Burney Phair with her model (left) of New Horizons Spacecraft with the "Venetia Burney Student Dust Counter" re-named in her honor in 2006. (Credit: NASA, Wikipedia, Martin George, 2009)

Hear a 1-17-06 podcast with NASA Public Affairs Edward Goldstein: **Venetia** shared her story the week prior to New Horizons' Launch: (https://www.nasa.gov/mp3/141071main_the_girl_who_named_pluto.mp3)

"I was having breakfast with my mother and my grandfather. And my grandfather read out at breakfast the great news and said he wondered what it would be called. And for some reason, I, after a short pause, said, '**Why not call it Pluto?**' I had read, and of course I did know about the solar system and the names the other planets have. And so, I suppose I just thought that this was a name that hadn't been used. And there it was. The rest was entirely my grandfather's work." **(Hear Venetia's voice yourself!)**

> **Lesson 11: Age Doesn't Matter! Venetia gave Pluto its name! The only girl in history to name a planet!** There are no limits to what you can do! **Nike**'s motto states: "Just Do it!"

Venetia was a well-read student in astronomy and planetary naming rules, thanks to her family connections. Her grandfather was **Falconer Madan**, a retired head of Oxford's **Bodleian Library**, and her **Great Uncle Henry Madan had, in 1877, suggested the names Phobos and Deimos, Greek gods of fear and terror for the newly discovered moons of Mars.** American astronomer **Asaph Hall** discovered the **two Mars' moons** using the 26-inch telescope at the U.S. Naval Observatory and liked the names which Henry Madan recommended. **Those are the names Asaph chose** because they are the two mythical sons of Ares, the god of war, also known by the Romans as Mars.

Venetia was simply following in Uncle Henry's footsteps in suggesting the name Pluto. Falconer knew and contacted Britain's former Royal Astronomer, now Oxford Professor Herbert Hall Turner.

On March 16, 1930, Turner sent a **telegram** (equivalent of today's email, but much slower) asking the astronomers at Lowell Observatory to consider the name "Pluto" for a dark gloomy planet, as suggested by a "small girl." The telegraph misspelled Venetia's name.

Side Note: Another famous Telegram sent by the Wright Brothers to their father on Dec 17, 1903 misspelled Orville's name. **See Library of Congress Wright Brothers Telegram** sent to their father Bishop Wright on Dec 17, 1903
https://www.loc.gov/item/mcc.061/

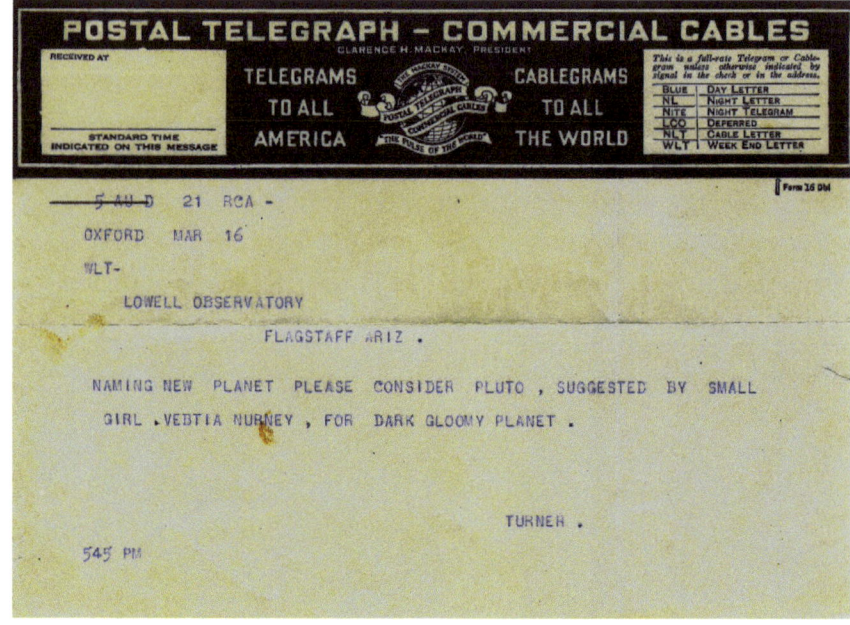

"Pluto" Telegram sent to Lowell Observatory, March 16, 1930
(Credit: Arizona Memory Project & Lowell Observatory Archives)

The Lowell staff liked the name Pluto because the 1st two letters of Pluto also honored Percival Lowell, who commissioned the search and spent 7 years searching unsuccessfully for Planet X (Pluto) at Lowell Observatory, which eventually hired Clyde.

The name Pluto was officially accepted by the IAU May 1, 1930.

Upon hearing that an 11-year old named the newly found world, **children across the globe took an immediate liking to little Pluto.**

Walt Disney seemed to take notice too. Mickey Mouse's pet dog appeared in a May 1931 cartoon called "The Moose Hunt". Mickey's dog's name? **Pluto the pup!**

Mythical Symbol for Pluto
(Credit: Wikipedia)

Venetia grew up to be an Accountant and a Teacher. In her later years, she received a model of the New Horizons Spacecraft as a gift from the Pluto Exploration Team. Venetia was also invited but did not attend the New Horizon's launch as a dignitary, on January 19, 2006. She said she valued her VIP Badge for the event. **And in her honor, the Student Dust Counter (SDC) aboard New Horizons' Spacecraft was renamed the "Venetia Burney Student Dust Counter".**

**Venetia remains the only girl to have named a planet! Way to go Venetia!
In 1987, Asteroid 6235 Burney was named in Venetia's honor as well.**

Venetia died at her home in Epsom England on April 30, 2009 at the age of 90.

> **Lesson 11.5: Pay no attention to your age.** Instead, pursue your dreams and goals and passions, and success will be yours. Note that the **Voyager** Records launched in 1977 contained a greeting by **6 year old Nick Sagan, Carl Sagan's** son. See p 71 & 92.

Features on Pluto photographed by New Horizons spacecraft in July 2015 were named for Venetia Burney, Percival Lowell, Clyde Tombaugh, and many others, as a way of honoring them. All names must be officially reviewed and accepted by the <u>International Astronomical Union</u>.

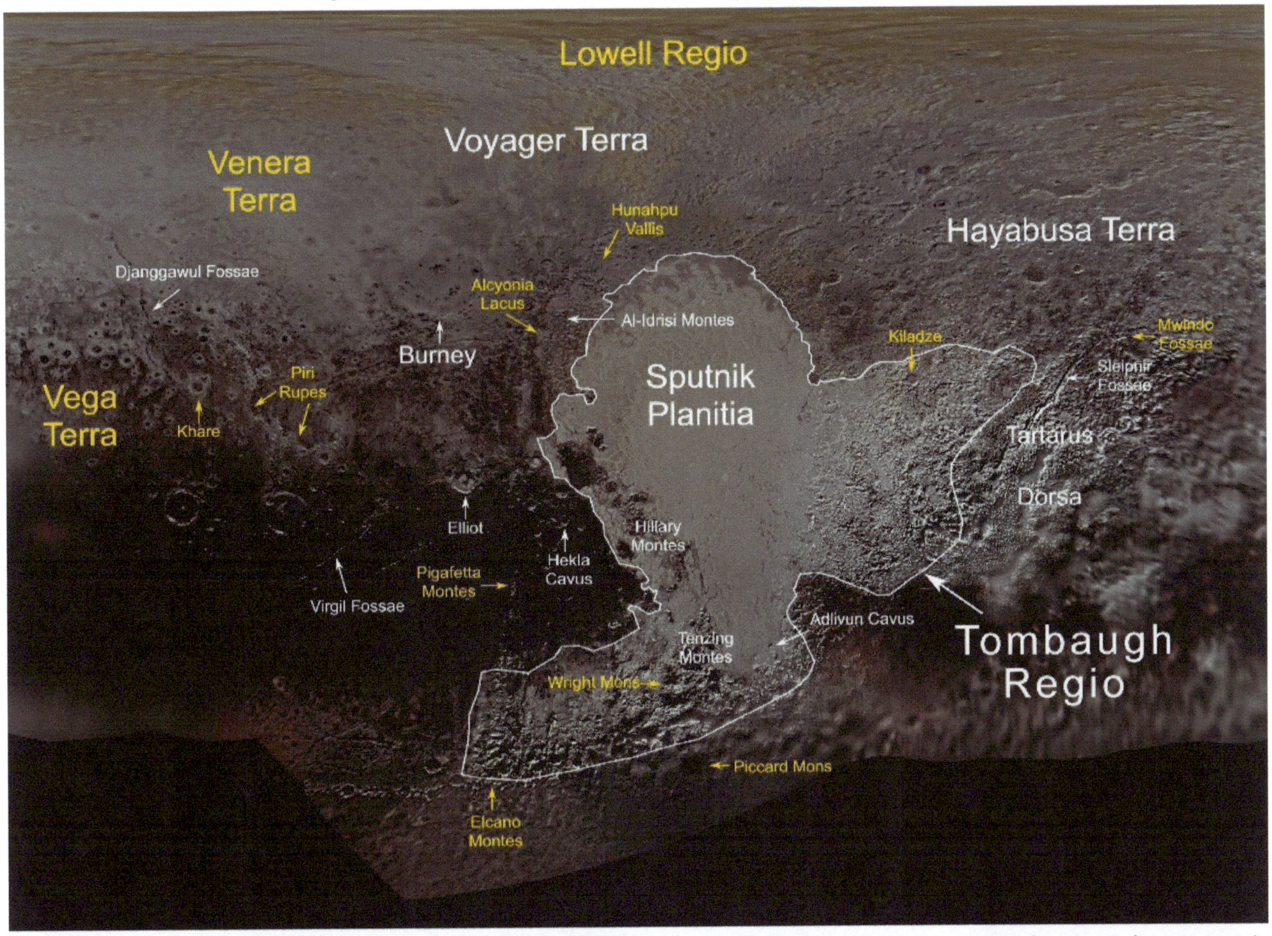

Pluto feature names officially approved by the <u>IAU</u> on Aug. 8, 2017 (Credit: NASA/JHUAPL/SwRI, IAU)

False-Color High-Resolution Images to enhance surface features, using RALPH Instrument Data (Credit: NASA/JHUAPL/SwRI/ZLDOYLE).

Let's have some fun with Age, Math, Fractions & Birthdays…on Pluto!

> **Lesson 12: Stay Young! It's an attitude! If people were born on Pluto, they'd spend their entire lives being less than 1 Pluto year old.** That's because Pluto's year (time it takes to orbit the sun) is 248 Earth years. So a 62 Earth year old would be 62/248 = ¼ **year old** in Pluto years! The people of Pluto would probably know a lot more fractions if they celebrated birthdays as often as we do on Earth (every 365 days). **So just for fun…Celebrate Fractional Birthdays! And Develop a Long Time Perspective!** Plan some long-term goals decades in the future!

In <u>Geometry</u>, <u>Navigation</u>, <u>Compasses</u>, and Cartography (map making), circles are divided into 360 equal degrees---see circle diagram below. Imagine Birthdays on Pluto. Because Pluto takes about 248 years to go around the Sun---this is 1 year on Pluto, to celebrate 1 Earth Year Old on Pluto, this would be 1/248 = 0.00403226 years old on Pluto. **So if you had a birthday cake on Pluto, imagine cutting a cake into 248 slices---1 slice per each Pluto fractional age.** See the top shaded slice of cake below, 360 degrees divided by 248 slices = 1.452 degrees per slice. We'd all be a lot skinnier for Plutonian Birthday Celebrations.

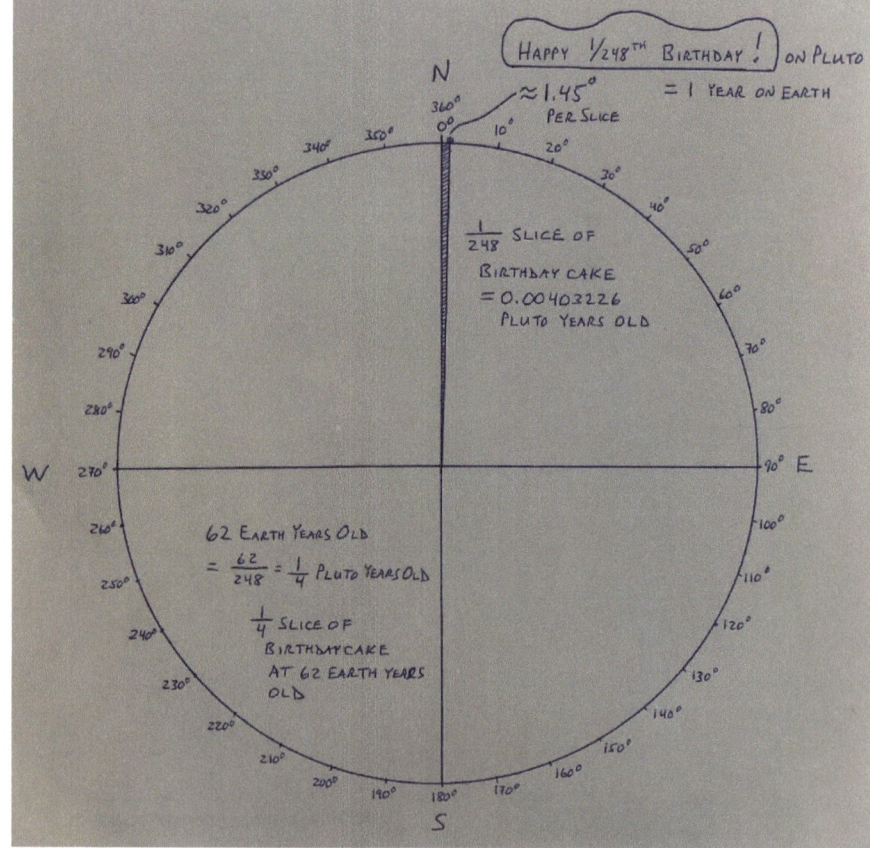

To say this another way, to reach your 1st Pluto Birthday you'd have to live to be 248 Earth years old! (Credit: NASA, https://spaceplace.nasa.gov)

(Right Credit: The Author)

And a 62-Earth-Year Old person on Pluto would be 62/248 or ¼ year old on Pluto---maybe they could have a ¼ slice of Birthday Cake (like the lower left quarter of the diagram).

Consider this: Pluto's Year (248 Earth years) is so long…That since the signing of the U.S. Declaration of Independence on July 4, 1776, Pluto has not yet traveled around the Sun once (as of this updated writing in January 2023)!

Chapter 3.7: You Just Discovered Pluto, Now What?

"Do not follow where the path may lead. Go instead where there is no path and leave a trail." – Ralph Waldo Emerson

In discovering Pluto, Clyde Tombaugh quickly became recognized as a world-class astronomer!

Clyde's hometown in Kansas was quite excited for him. His former school principal worked to get Clyde a 4-year scholarship at the **University of Kansas** in 1931, at last able to obtain the formal astronomy education he desired! At 26 years old, Clyde enrolled in Fall, 1932. The Kansas City Star newspaper published a photo of Clyde, and the U.S. Postal Service even delivered letters addressed to: Clyde Tombaugh, Kansas. Clyde's good fortune continued while at the University of Kansas: Clyde met his future wife **Patricia Edson**.

Clyde Tombaugh in his later years with Hubble Telescope View of Pluto and largest moon Charon (left photo) and Clyde in his back yard in 1996 with one of his home-built Telescopes (above). (Credits: NASA, https://www.gsfc.nasa.gov and (above) Sky&Telescope, J. Kelly Beatty, 1996)

In his book **"Clyde Tombaugh, Discoverer of planet Pluto"**, astronomer-author **David H. Levy** describes an introductory astronomy class for which Clyde signed up:

> The Chairman of the department which ran the astronomy class, **Dr. Dinsmore Alter**, learned about Clyde Tombaugh signing up for the class. "He would have none of it. The idea of the discoverer of the solar system's ninth major planet sitting happily in the introductory astronomy lecture hall and answering questions about **Kepler's laws** seemed ludicrous to him." Clyde was disappointed because he thought it would be "fun just to go through it."

Clyde graduated in 1936 and in 1938 started his Masters degree as well, graduating in 1939.

> **Lesson 12.5: Become a Life-Long Learner.** This is a minimum requirement for success in the 21st century and beyond.

The United States entered World War II with the attack on **Pearl Harbor** on December 7, 1941 and Clyde became the commander of Civil Defense in Flagstaff's Coconino County. In 1943 he was called to serve and taught Navigation to the United States Navy at **Arizona State College** from 1943-1945.
In 1944 Clyde was also invited to teach astronomy at the University of California, Los Angeles.

Clyde had taken a leave of absence from the Lowell Observatory for military service, to attend school and to teach at University. **He had planned to return to Lowell following military service, but**

Dr. Slipher informed Clyde there were no funds to re-hire him---Clyde was being let go after 14 years! In his book, **"Out of the Darkness: The Planet Pluto"** Clyde estimates he had reviewed **over 90 million star images during 7000 hours at the Blink Comparator. Now he was being dismissed!**

In his book, **"Clyde Tombaugh...Discoverer of Planet Pluto"**, author-astronomer **David H. Levy** relates this story:

> **Clyde's wife Patricia (Patsy) said,** "He was so hurt he didn't even tell me."
>
> And in Clyde's own words, "I felt I had earned a status of some import and they weren't going to give it to me...**I felt terribly cheated.** I had all these years of hard work, hard blinking." [Important Author's Note: Clyde re-grouped and moved forward!]
>
> Important work on galaxy distribution was abandoned---Clyde had photographed 29,548 galaxies, discussed with **Edwin Hubble** the irregularities and voids in the galaxy distribution throughout the Universe, and so much more.

From 1946-1955, Clyde worked as an optical physicist, astronomer and ordnance engineer responsible for a team of 75 people at the White Sands Missile Range in Las Cruces, New Mexico to track captured German V2 rockets. After 9 years of service, Clyde earned the Medal of the Pioneers of **White Sands Missile Range**. From 1958-1973, Clyde taught Astronomy at New Mexico State University and in 1960 was awarded an honorary Doctor of Astronomy Degree from Northern Arizona University.

Still image from a video of personal interview with Clyde Tombaugh in his backyard home in Las Cruces, NM on Oct 26, 1991.

Clyde explained that this is his 9-inch home-built Newtonian reflecting telescope (made with discarded farm and car parts).

This is the telescope through which he saw Jupiter and Mars, then made sketches which he sent to Lowell Observatory, **which led** to his being hired for the Planet X search, **which led** to his discovery of Pluto, **then** college scholarship, **meeting** his wife, **having a** family & a **successful career. Wow! What a sequence of events!**

Credit: ("Clyde Tombaugh, Discoverer of the Planet Pluto – **Academy of Achievement**," 2019).

Visit https://achievement.org/achiever/clyde-tombaugh/ for Clyde's interview. AND, this site also features: **Sally Ride, Alan B. Shepard Jr, Chuck Yeager, Story Musgrave, Charles Townes and more!**

In his book "Clyde Tombaugh, Discoverer of Planet Pluto", author-astronomer David H. Levy says:
"During Clyde's research at Lowell Observatory, he photographed the following astronomical objects:

- 3969 Asteroids (775 possibly never observed previously)
- 2 comets
- 1 nova
- 5 open star clusters
- Several clusters of galaxies
- 1 supercluster of galaxies (Clyde counted 29,548 galaxies in all!)
- 1 globular star cluster
- **1 trans-Neptunian Object: Planet Pluto (1930), "dwarf-planet" Pluto (since 2006)!**
  Wow! What an impressive legacy of work! Clyde retired in 1973.

> "Well Done is better than Well Said."
> – **Benjamin Franklin**

Let's digress to discuss Two Amazing Achievements:

1. On February 18, 2021, **Perseverance Rover** landed on Mars
See https://www.youtube.com/watch?v=M4tdMR5HLtg
2. On April 19, 2021, **Ingenuity Helicopter made the 1st powered flight on another planet,** rising 3 meters, hovering, pivoting, landing! Because Mars' atmosphere is so thin, Ingenuity's propellers (with a 4-foot wingspan) need to spin 2500-2800 Rotations Per Minute (RPMs) versus commercial helicopter blades on Earth at 400-500 RPMs.

Ingenuity carried a small piece of fabric from the Wright Brothers' aircraft to Mars **to celebrate this achievement! The Wright Brothers would be proud: 118 years from 1st Powered flight on Earth to 1st powered flight on Mars!**

Mars Helicopter **Ingenuity** (4 lbs, 1.8 Kg) and **Perseverance** Rover on Mars, April 2021 (Credit: NASA/JPL)
4-13-23: Ingenuity made its 50th flight!

Mars Helicopter Ingenuity (left) and its 2nd Powered Flight on another planet 4-22-21 (right). Photos by Rover Perseverance (Credits: **NASA/JPL** – Caltech/**Malin Space Science Systems**)

Watch Ingenuity's 1st Flight on Mars here: https://youtu.be/wMnOo2zcjXA (video only)
and Exciting NASA JPL Team Story as it happened, here: https://youtu.be/ia6S1jZmwWc

On December 10, 2022, Ingenuity made its 36th flight on Mars, **far exceeding proof of concept goals! In 2027, NASA will launch "Dragonfly", a heavier nuclear-powered 8-rotor Drone to explore Saturn's moon Titan, with arrival in 2034!** It will take-off, land and explore---No Batteries Required! Why not?

NASA's Jet Propulsion Laboratory motto fits well: "Dare Mighty Things!" (See www.jpl.nasa.gov)

And now back to Clyde:

In August 1992, Jet Propulsion Laboratory Scientist Robert Staehle called Clyde Tombaugh and requested Clyde's permission to visit "his planet" with a robotic mission to explore Pluto. Clyde gladly agreed. A series of proposals to NASA followed which eventually became the New Horizons Flyby Mission to Pluto.

Clyde became the 1st person to have his ashes sent to the edge of the solar system and beyond.
(Credit: NASA, **Johns Hopkins University Applied Physics Laboratory**, **Southwest Research Institute**)

Clyde died at age 90 at his home in Las Cruces, New Mexico on January 17, 1997. Prior to the New Horizons launch in 2006 and to honor Clyde further, Alan Stern, the principal investigator and leader of the New Horizon's mission to Pluto asked Clyde's widow if she'd be willing to donate some of Clyde's ashes to be interned within a 2-inch aluminum container to be attached to the New Horizons Spacecraft. Alan Stern wrote the dedication message for the long voyage. Clyde's wife Patsy witnessed the launch of the New Horizons spacecraft on January 19, 2006.

> **The Discoverer of Pluto thus became the first person whose ashes have travelled to Pluto and beyond into the outer limits of our Solar System: Into the Kuiper Belt.**

For some beautiful STEAM Artwork: View the stained-glass tribute to Clyde at the Unitarian Universalist Church of Las Cruces New Mexico which Clyde and Patsy Tombaugh helped found in 1955. **Artist** Arthur J. Tatkoski created the 8 foot by 18 foot glass mural---dedicated in 2001---depicting and honoring founder Clyde Tombaugh's Scientific Achievements:
Google: "Clyde Tombaugh Stained Glass Window"

Chapter 4: James Christy, Honoring Mrs. Christy

"A lot of husbands promise their wives the Moon, but he delivered!"
- ### Char (Charlene) Christy, Jim's wife

In June of 1978 **James (Jim) Christy** was working at the **U.S. Naval Observatory** (which has a 61-inch telescope) in Flagstaff, about 6 miles from where Pluto was discovered. He went to his supervisor **Robert (Bob) Harrington** and asked if he had any simple projects he could work on. Bob pulled out an envelope from his desk drawer containing photographic slides of Pluto taken by the Lowell Observatory, and gave the envelope to Jim to review.

Jim took the slides to a microscope to examine the images and noticed an **asymmetry,** elongation or a bump on the images of Pluto. It was as if Pluto was out of focus. He checked other slides of Pluto and compared them to star images on the same film. **The stars images were not elongated but focused, and Pluto's elongation or bump had moved to a different orientation or disappeared altogether.** Either Pluto had a large mountain or a moon orbiting very close by.

Jim concluded that Pluto has a moon!

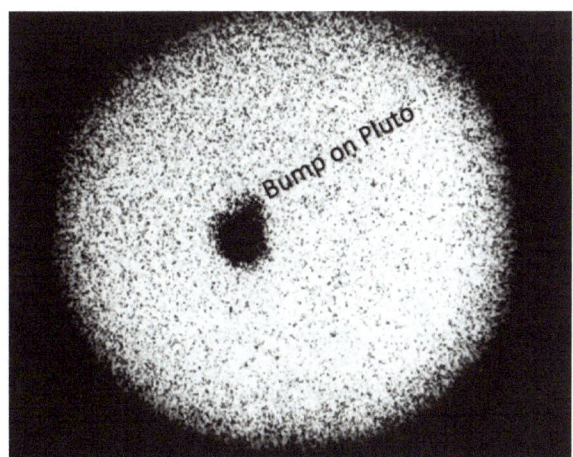

Credit: U.S. Naval Observatory   Credit: U.S. Naval Observatory.

In a documentary film celebrating the anniversary of discovery, Jim's wife Char related this story:
Jim called me and said, "I think I'm gonna be famous. I think I've discovered a moon!"

Then Jim went to Bob and said, "Pluto has a moon!" Bob replied, "Jim, you're crazy!"

Together Jim and Bob reviewed the evidence, cross-checked, calculated the timing, and took additional photographs of Pluto using the 61-inch telescope. They verified their work with previous images. This helped them calculate Pluto's rotation rate was 6.39 Earth days---or the length of 1 day on Pluto.

On July 2, 1978 they confirmed that Pluto had a large close moon. This meant Pluto was a **"double-planet"** system with Pluto and its Moon orbiting a common center of mass. **They announced the discovery to the world on July 7, 1978.** Bob is often credited as the co-discoverer for his efforts too.

As discoverer of the new moon, Jim was offered the opportunity to name it. He said he wanted to name the Moon after his wife Char (short for Charlene). Jim said that he often thought about protons and electrons, so he added the "-on" to Char and suggested "Charon".

Charlene tells the story that Jim spent a sleepless night worrying whether his recommended name would be formally accepted by the International Astronomical Union, which oversees the naming of astronomical objects, **because they usually only approve names from mythology.** So he got up that night, grabbed a dictionary, and looked to see if the word "Charon" was there…hoping it would be!

To his great relief, there in bold print, he found the word "Charon", a character in Greek mythology who ferries the souls of the dead over the river Styx:

> **Cha·ro·lais** \ˌshar-ə-ˈlā\ n [*Charolais*, district in eastern France] (1893) : any of a breed of large white cattle developed in France and used primarily for beef and crossbreeding
> **Char·on** \ˈkar-ən, ˈker-, -än\ n [L, fr. Gk *Charōn*] : a son of Erebus who in Greek mythology ferries the souls of the dead over the Styx
> **char·poy** \ˈchär-ˌpȯi\ n, pl **charpoys** [Hindi *cārpāī*] (1845) : a bed used esp. in India consisting of a frame strung with tapes or light rope
> **charr** \ˈchär\ var of ¹CHAR

(Credit: Kevin Caruso, Merriam Webster's Collegiate Dictionary, 10th Edition)

The International Astronomical Union accepted Jim's recommended name, and Char was honored.

With a smile, Jim's wife kiddingly said on the documentary of the 40th anniversary of Charon discovery:
"A lot of husbands promise their wives the Moon, but he delivered!"

Hear Jim Christy, Randy Monroe (Jim's stepson) and Char Christy tell their story in this NASA video:
https://youtu.be/5zmETUjRXzM

Astronomer James Christy with his now famous discovery images of Pluto's largest Moon Charon.
(Credits: James Christy and U.S. Naval Observatory)

> **Lesson 13: Be Thorough. Give your work---even a simple task---the serious care and attention it deserves.** Stay alert and observant. You may be on the brink of an amazing discovery! **And Be a Champion (it's an attitude)**---optimistic, curious, creative, forward-focused, ever-learning, and open to new discoveries.

Time Out. Let's look at Great Undersea Explorer: Jacques Cousteau

"We must go and see for ourselves." – Jacques Yves Cousteau

Jacques Cousteau (1910-1997): French Naval Officer, Self-Taught Oceanographer, Self-Taught Film Maker and Director, Author, Ocean Life Conservationist, Co-inventor/Co-Patent-holder of the Aqua-Lung SCUBA (Self-Contained Underwater Breathing Apparatus) Regulator.

Cousteau wearing his Aqua-Lung
(Credit: www.pinterest.com)

Water Spider "Aqua-Lung"

- Learned to swim age 4 (Credit: Infographic4U.com)
- Learned underwater diving at summer camp
- Enjoyed understanding mechanical devices
- **Wanted to become a Pilot** in the French Navy
- Joined French Naval Academy after college
- Crashed his father's car in **near-fatal accident** in 1933, **breaking both arms**
- Rehabilitated his arms **by swimming daily** in the Mediterranean Sea
- Modified Fernez underwater goggles in 1936 to see clearly underwater
- Modified camera equipment for underwater photography
- Fascinated How Water Spiders carry an Air Bubble to breathe underwater
- Filmed an underwater movie titled "18 Meters Deep" with 2 friends, which won praise and an award
- **Studied early designs of SCUBA** (Self-Contained Underwater Breathing Apparatus) invented mid 1920s which required a heavy metal helmet, diving suit and tethered air hose attached to a surface ship in order for the diver to breathe underwater
- **Studied improved but not yet perfected SCUBA devices** which required a diver to carry an air tank on their chest and **manually turn a value on the tank when they needed to breathe,** which wasted tank air
- Worked with Engineer Emile Gagnan in 1937 and together modified an automotive valve to create a safe chest-mounted air regulator for divers so air would be fed to the diver on-demand when they breathed-in from hoses connected to air tanks on a diver's back---freeing up hands, conserving the air so divers could swim independently (untethered) of a surface ship and remain underwater longer
- **Co-invented and Patented their Aqua-Lung Diving Unit**---now part of virtually every modern open-circuit SCUBA (Self-Contained Underwater Breathing Apparatus), allowing divers world-wide to swim and breathe freely as Aquanauts: undersea explorers. Google: U.S. Patent No. 2,485,039 filed on 3-10-47
- Explored Earth's Oceans from his research vessel "Calypso", including a 13,800 mile voyage in 1955
  Note: In 2019, NASA's new Crew Space Transportation CST-100 spacecraft was named "Calypso"!
- Won acclaim and awards for his 90-minute Television Documentary "The Silent World" in 1957
- Awarded the Gold Medal of the National Geographic Society by President John F. Kennedy in 1961
- Maintained detailed log books, Invented: underwater 35mm camera, Diving Saucer (small submarine), Shark Cage, 3 undersea habitats (Conshelf) which proved people could live underwater for long periods AND discovered the necessity of sunlight for human health. See film "World Without Sun"
- Produced 115 television films and 50 books on oceanography, marine life and the importance of ocean and marine life conservation, and was inducted into the Television Academy Hall of Fame in 1987.
- Established in 1973 "Cousteau Society" for conservation of our oceans/ocean life. www.Cousteau.org
- Received the presidential medal of freedom by President Ronald Reagan in 1985

See www.Cousteau.com re: Jacques' explorations, inventions, ongoing legacy. And, Google: Water Spiders!

Cousteau's life and attitude are shining examples of overcoming a life-altering set-back (near fatal car accident) & choosing to re-direct his energy and self-learning to a new life-long passion (oceanography, conservationism, documentaries) for the benefit of the people of Earth.
Like Lowell, Clyde & the Wrights, Jacques developed self-taught credentials to become a world-class expert!
In what subjects are you passionate? How do you want to excel & share your talents with the world?

Chapter 5: Hubble Space Telescope, 5 Moons!

"Life is either a daring adventure or nothing at all." – Helen Keller

Astronomers didn't get a better view of Pluto until the **Hubble Space Telescope** was released into Earth Orbit from the payload bay of the Space Shuttle Discovery.

The Hubble Space Telescope was deployed into Earth Orbit from the cargo bay of Space Shuttle Discovery on April 24, 1990. (Credit: NASA)

Best view of Pluto and Charon as Seen by the Hubble Space Telescope on Feb 21, 1994.
(Credit: Dr. R. Albrecht, ESA/ESO Space Telescope European Coordinating Facility; NASA)

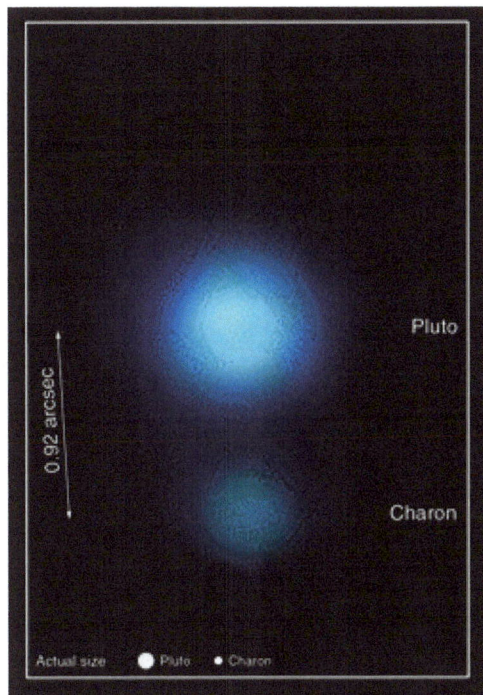

If you visit the Big Island of Hawaii, I recommend visiting the town of Hilo, and the **impressive Imiloa Astronomy Center.** On a clear day, look for telescopes atop Mauna Kea!

If you're in good health, drive to the Subaru Telescope Hilo Base Facility and ASK IF you can Schedule a "Summit Tour" of Subaru Telescope (1 of 13 building-sized telescopes atop Mauna Kea, near the 13,796-foot peak). Subaru is the Japanese word for the **Pleiades star cluster**, and the stars are the Subaru Logo. The Hawaiian word is **"Makali'I"**. **Caution: There's Much Less Oxygen atop Mauna Kea, so follow safety instructions & acclimate if you do visit.**

Best view of Pluto and Charon (in **infrared light**) by the 8.2 Meter Subaru Telescope atop Mauna Kea mountain near Hilo Hawaii. (Credit: **Subaru Telescope, National Astronomical Observatory of Japan**) Discoveries, 3D Tour, Subaru Hawaii Careers: https://subarutelescope.org/en/

Ever consider a career in Astronomy in Hawaii?

Lesson 14: Keep your goals in sight, keep making progress and your vision will become clearer.

By blocking out brighter Pluto and Charon (center vertical black stripe below), and taking ever-longer exposures, scientists operating the Hubble Space Telescope over a period of several years were able to resolve & discover that **Pluto has 5 moons!** **Nix** and **Hydra** were discovered in 2005, **Styx** in 2011, and **Kerberos** in 2012. WOW! Pluto was becoming even more interesting...its own system!

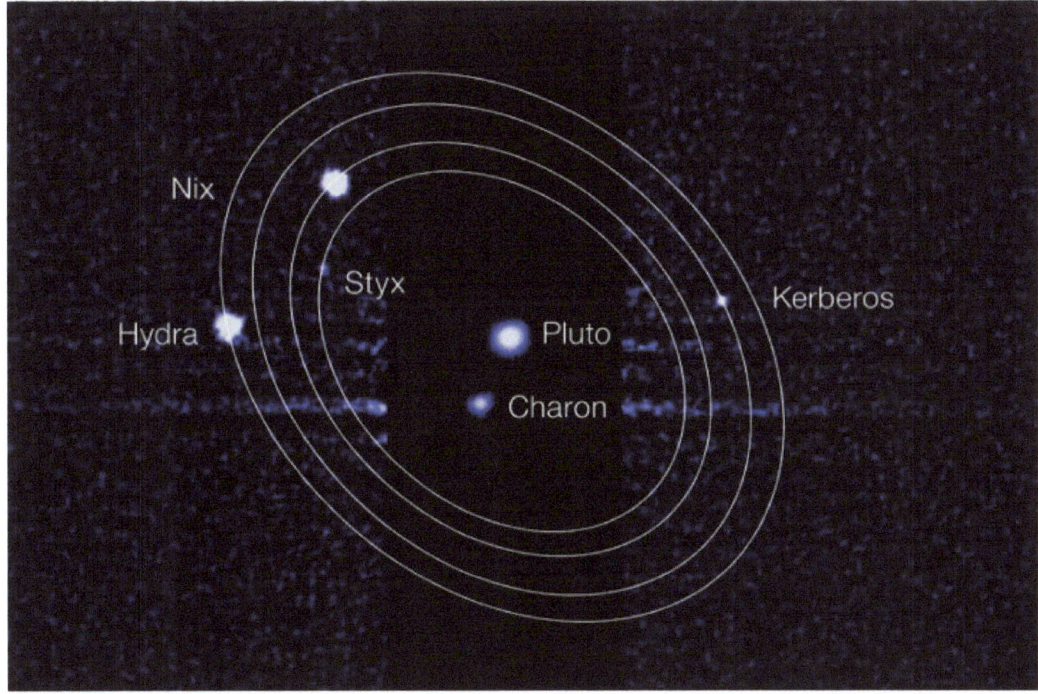

The Pluto System: Pluto and its 5 Moons (orbital paths added for clarity).
(Credit: NASA, ESA, and **Mark Showalter** (SETI)). See www.seti.org/our-scientists/mark-showalter)
Astronomer Mark Showalter discovered two of Pluto's moons using Hubble: Styx and Kerberos!

Continuous Improvement: Clyde developed a love for reading and learning...

A 2021 Gallup Survey of Americans on their reading habits found that:

American Adults read 12.6 books in 2021; but 17% of Americans read NO books in 2021. Women read 15.7 books compared to Men who read only 9.5 books in 2021. (that's all forms: ebooks, Printed, Audiobooks)

Nonfiction books represent compressed time, often 10-40 years of the author's life experience summarized in a book, for a few dollars.

What does that mean to you? It means that you can buy books for a few dollars, read them, and gain insights from the authors' 10-40 years of experience---all without having to spend 10-40 years of your own life learning it yourself! Wow! **Can you see the advantage this provides you in knowledge, and in the job market as well? This is Compressed Time!** This means you can put yourself at the front of the knowledge line whether you attend formal school or not. How many books have you read this year? Get started today.

This means that you can leverage the experience of others and save yourself years to decades of your own life! I recommend exceeding the Gallup average and reading **2 books per month or 1 hour per day** in your field of interest, just 5 days a week, times 50 weeks a year **or 250 hours per year.** Divide by 40 hours/week: **That comes to 6.25 forty-hour work weeks/year!**

That's 6.25 forty-hour work weeks per year, reading & learning in your field of interest.
Do you see how simple is it to become an expert in your chosen field and rise above most of your competition? What would happen if you dedicated 6.25 work weeks per year to learning about your work or area of interest? **Do you think that might influence your income? As we said in the last chapter, if you're not taking in new ideas regularly, you are moving backwards, compared to the world, which is constantly changing, growing, improving.**

Next: The American Automobile Association (AAA) 2020-2021 Driving Survey found:
Americans (16 years old and up) drive 61.3 minutes per day. Multiply by 365 days and divide by 60 minutes per hour and that comes to 372.9 hours per year. Now Divide by 40 hours/week: **That's 9.3 forty-hour work weeks behind the wheel in 2021.**

What if you took 9.3 work weeks per year to study your field of work or interest? Would that have a small or large impact on your income? LARGE! Do you see the positive benefits on your life? You have complete control over this! Just decide and get started learning!

Key Insight: Knowledge, Change, Global Collaboration & Competition are Accelerating! This means, You Cannot Stand Still! After graduation, saying "I'm done learning" is a **monumental mistake! Your School Education WON'T LAST a lifetime! Think of your knowledge as a leaky bucket.** If you don't keep filling, you'll be moving backwards because your peers & competition will continue to fill & advance. **Continuous Improvement & learning in your chosen field is a Minimum Requirement to increasing your value! So? In what field(s) do you want to excel?**

Chapter 6: Alan Stern, 29-Cent Challenge...Pluto Explored!

President <u>John F. Kennedy</u> asked rocket scientist <u>Wernher von Braun</u> (who, afterward, designed the <u>Saturn V</u> rocket which took <u>Apollo 11</u> astronauts to the Moon), "What will it take to build a rocket to take a man to the moon and safely bring him back to Earth?" Wernher replied, "The will to do it, Mr. President!"

On October 1, 1991, the United States Postal Service issued a series of Solar System Exploration Stamps for the planets. The stamp for Pluto stated: "Pluto Not Yet Explored" (Credit: U.S. Postal Service, and the author)

New Horizons Spacecraft
(Credit: NASA/<u>JHUAPL</u>/<u>SwRI</u>)

In his book "<u>Chasing New Horizons</u>, inside the epic first mission to Pluto", scientist-author, and principal investigator of the New Horizons mission <u>Alan Stern</u> said **"<u>Rob Stachie</u> and <u>Stacy Winstein</u>** [two engineers at the Jet Propulsion Laboratory] took this stamp as a dare" and motivation to present a plan which NASA would accept, to finally explore this far-away world.

Six Times, they proposed mission plans to NASA to explore Pluto, and six times were told "No!"

What do YOU do when you're told "No!" and you really believe in your idea? Do you give up? NO! Expect to hear "NO!" when sharing your important ideas and dreams and goals with others---it's very normal and part of the achievement process. Just stay positive and focused and gather additional information in order to learn how to keep moving in the direction of your goals.

Did the Pluto exploration team give up? NO! They kept going. They persisted. They rewrote and re-proposed their plans repeatedly. **Finally, NASA said YES to funding the mission!**

It took 4 more years to build and launch, plus 9.5 more years to track and monitor the flight to Pluto.

Stern and 11 subject matter experts about Pluto (they called themselves the **<u>Pluto Underground</u>**) first met in a restaurant in 1989 to discuss pushing NASA to explore little Pluto. From that meeting in 1989 to the 2015 successful Pluto Fly-by, **26 years elapsed---that's persistence!** At launch, Pluto was a Planet, and during the 9.5-year flight to Pluto, it changed to a "Dwarf-Planet". **Another example of <u>persistence, tenacity, and perseverance.</u>**

Let's digress again. This time, Let's talk about Ants! Ants? Yep!

In his book entitled "7 Strategies for Wealth & Happiness", author, speaker and business philosopher **Jim Rohn describes "the Ant Philosophy".**

When was the last time you watched an ant walking on the ground?
Did you ever try to block its path?
What did it do?
Did it stop? You know the answer! NO!
The ant searched for another way: Over, under, around, back, sideways.

Did it quit? You know that answer too! NO!
It kept looking for another way.
For how long?
UNTIL it found another way!

(Credit: the Author)

What a great word: UNTIL!
Ants Don't Quit!
What a Great Philosophy!

Jim further says: "Guess what ants plan for all summer? That's right---winter. And how much will an ant gather in the summer to prepare for winter? *All* it can! How intelligent!"

What's the moral of the story?

Pursue your goals UNTIL you achieve them. Find a Way. Don't Quit. During the Summers of Life, Prepare for the Winters. During the Winters of Life, know that Spring is on the way! Follow the Ant Philosophy in your life!

When obstacles appear to block your path, Find Another Way! Don't Quit! Never Give Up! Keep Looking for Another Way! Go under, over, around, back, sideways (OR THROUGH!) UNTIL you find another way! That word "UNTIL" means you'll never stop looking for another way.

Lesson 15: Expect Adversity & Obstacles on the path to your goals. Adapt. Push Forward.
When you set out to accomplish something which has never been done before, **obstacles WILL appear (not may appear, but WILL appear!)** and **people will tell you "NO!" many times.** If the goal is worthy and important, and it doesn't harm any other person, then don't let any obstacles stop you! Use obstacles as steppingstones in the direction of your goals!

So far, we've mentioned water spiders (breathing underwater) and ants (not quitting).
What other examples from Nature can you think about from which people might be able to benefit?
Peek ahead to **The Law of Cause and Effect** (pages 39-40) for several more examples which help people.

New Horizons Principal Investigator **Alan Stern** plus **Applied Physics Laboratory** director Ralph Semmel (middle) plus Clyde's daughter Annette and son-in-law Wilbur proudly display an image of the former postage stamp with the words "NOT YET" crossed out, now reading: Pluto Explored, in July 2015. (Credit: NASA, Johns Hopkins University Applied Physics Laboratory, Southwest Research Institute, Bill Ingalls). **Goal Achieved! The "Original 9 Planets" were now Explored!**

Goal Achieved! 26 years after the initial meeting to discuss exploring Pluto!

In his book "The Case for Pluto", author **Alan Boyle** quotes Alan Stern regarding flying the stamp aboard the New Horizons spacecraft,

"I wanted to fly it [the stamp] as a sort of 'in your face' thing."

Bravo Alan Stern! Sometimes a little drama is needed!

"When the dream is big enough, the odds don't matter!" – Dexter Yager

"Never give up. Never, Never give up!" – Sir Winston Churchill

"There ain't no rules around here. We're trying to accomplish something." – Thomas Edison

"Make "NO!" your vitamin!" – Les Brown

"It takes sixty-five thousand errors before you are qualified to make a rocket." – Wernher von Braun

---

Lesson 16: The process of scientific discovery is an unending, fluid flow of change and learning.

But Wait! New Horizons wasn't done yet!...

Consider this for a few moments... The New Horizons spacecraft was launched on January 19, 2006. Its mission was to explore the Pluto System and, as an extended mission option, the Kuiper Belt beyond.

January 19, 2006 Launch of New Horizons spacecraft on a journey to an empty place in space where Pluto would be in 9.5 years!
(Credit: NASA)
See Page 27 NASA video Link about Charon which Also shows the launch Video of New Horizons).
https://youtu.be/5zmETUjRXzM

The laws of physics are so precise that we can launch a spacecraft to Pluto, aiming at an empty point in space where Pluto and the spacecraft will arrive together 9.5 years in the future, and time that rendezvous within a few minutes! Wow! Simply amazing isn't it?

Because sunlight at Pluto is about 1000 times dimmer than at Earth, the **New Horizons spacecraft could not use solar panels for power.** Because the 1-way journey would take about 10 years, **New Horizons could not be battery powered**---they would be long dead in the extreme cold of space. However, NASA has been using **radioisotope thermoelectric generators (RTGs)** in space for over 50 years.

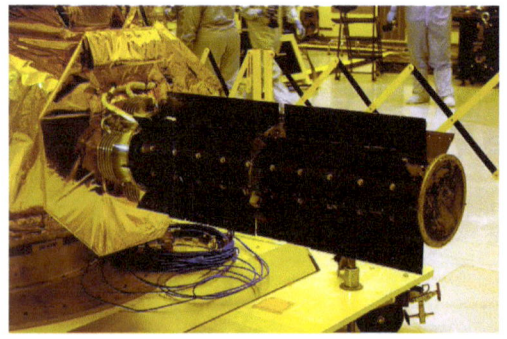

New Horizons spacecraft power source: Radioisotope Thermoelectric Generator (Credit: NASA, JHUAPL, SwRI)

RTGs have no moving parts---instead, **radioactive plutonium dioxide** fuel decays naturally and emits heat. A thermoelectric converter changes the large temperature difference between the decaying fuel and the cold vacuum of space to electricity, and the excess heat keeps the electronics warm enough to operate. (Photo Credit: NASA, JHUAPL, SwRI)

In the 2015 hit movie, "The Martian" (book by **Andy Weir**, directed by **Ridley Scott**), stranded astronaut **Mark Watney** (played by actor **Matt Damon**) digs up their **RTG** for Warmth. See the 2015 YouTube Movie Trailer for "The Martian":
https://www.youtube.com/watch?v=Ue4PCl0NamI

Chapter 7: IAU Astronomers, Defining "Planet"

"There can be no thought of finishing, for 'aiming at the stars', both literally and figuratively, is a problem to occupy generations, so that no matter how much progress one makes, there is always the thrill of just beginning."
– Dr. Robert Goddard

Dwarf-Planet Ceres in False-Color with bright material in Occator Crater (Credit: NASA/JPL-Caltech /UCLA/MPS/DLR/IDA) 12-9-2015 photo – Dawn probe.

The asteroid belt was unknown on January 1, 1801, when Giuseppe Piazzi was making a star map and discovered Ceres, the largest object within the asteroid belt between the orbits of Mars and Jupiter. Giuseppe was an Italian Catholic priest, mathematician and astronomer who established the Palermo Astronomical Observatory in Italy.

Ceres was 1st thought to be a planet, but as many more objects were later discovered between Mars and Jupiter, it became clear that Ceres was an asteroid, the largest in the asteroid belt. In August 2006, the IAU re-classified Ceres as a Dwarf Planet---the only dwarf-planet which remains within Neptune's orbit. NASA's "Dawn" spacecraft orbited Ceres from 3-6-15 through 10-31-18, when its fuel ran out. Ceres is named after the Roman goddess of grain crops/harvest. Note: The word "cereal" has the same origin. See https://www.youtube.com/watch?v=PRgpuvjV_Vs

Ceres is an example of the dynamic and exciting process of Science and the Scientific Method---as astronomers improved their understanding through improved data to better fit their observations, their theories and classifications changed: Ceres' classification went from planet to asteroid to dwarf-planet.

Years after Clyde Tombaugh discovered Pluto in 1930, astronomers speculated that there might be more objects beyond Neptune. In 1943, Irish Astronomer Kenneth Edgeworth suggested in published papers that there could be many small objects in the region beyond Pluto. In 1951, Dutch Astronomer Gerard Kuiper suggested there might be many objects in a zone beyond Neptune, but that Pluto's gravity may have already scattered them to a region much farther away, named the Oort Cloud.

The development of Charge-Coupled-Devices (CCDs) provided a new tool for astronomers in computer guided sky searches versus having to depend upon their eyes looking at photographic plates. In 1986, Astronomers David Jewitt and Jane Luu launched a "slow-moving object" sky survey. They outfitted a 2.2-meter telescope atop Hawaii's Mauna Kea mountain with a CCD and began searching for objects beyond Saturn. They leveraged CCD technology and "digital blinking" not available to Clyde Tombaugh.

After 6 years of searching, on August 30, 1992 they discovered and cataloged 1992QB1 (later named Albion)---the first Kuiper Belt Object beyond Pluto---a Trans Neptunian Object (TNO), thereby confirming the existence of the Kuiper Belt. By definition and location, Pluto was first TNO or Kuiper Belt Object. On 3-28-93 they discovered KBO 1993FW and on 9-14-93 a third KBO: 1993RO.

In his **book "The Pluto Files"**, Astronomer and author **Neil de Grasse Tyson** describes a panel debate about Pluto that he organized at the American Museum of Natural History in NY on **May 24, 1999** entitled: "Pluto's Last Stand: A Panel of Experts Discuss and Debate the Classification of the Solar System's Smallest Planet."

About 800 subject matter experts from many countries participated, including: Jane Luu (co-discoverer of the 1st Kuiper Belt object beyond Pluto), David Levy (biographer of Clyde Tombaugh and discoverer/co-discoverer of many comets and asteroids), Alan Stern (expert in "all things small in our solar system" and future principal investigator for the future New Horizons Mission to Pluto).

- **Jane Luu reminded her peers that Ceres had its planethood revoked** when it was found to be the largest member of a new class of objects in the solar system: Asteroids in the Asteroid Belt.
- **Alan Stern suggested** a physical mass test for planethood: An upper limit less massive than a star and a lower limit of massive enough to achieve hydrostatic equilibrium (a sphere shape).
- **David Levy emphasized that science is for everyone, not just scientists. He suggested sending a spacecraft to photograph Pluto** and see if it looks like a planet. For now, leave Pluto alone.

Founded on July 28, 1919, the **International Astronomical Union (IAU)** is a non-governmental international association of professional astronomers at the PhD level and beyond, active in professional research and education in astronomy. The IAU is based in Paris France and comprises 13,701 members (as of 2019) from 102 countries **whose mission is to advance scientific progress in astronomy, to promote astronomical research, outreach, education, provide a forum for scientific debate and development through global cooperation.** They're also responsible for naming, defining and classifying astronomical objects. Science has the challenging job of assigning classifications or "boxes" to what we see in nature. This is a dynamic and evolving process. Nature doesn't always fit neatly into our "boxes".

Using advancing technology and automated search software, many more KBOs were soon discovered:
- KBO 2000 WR106 (later cataloged **Varuna**) was discovered 11-28-00.
- KBO 2002 LM60 (later cataloged as **Quaoar---about ½ Pluto diameter**) was discovered 6-4-02.
- KBO 2003 UB12 (later cataloged as **Sedna---about ¾ Pluto diameter**) was discovered 11-14-03. Sedna's discovery caused the IAU to form a 19-member panel in 2005 to help define "planet".
- KBO 2003EL61 (later cataloged as **Haumea** was discovered 3-7-03.
- **KBO 2003 UB313 (first named Xena and later cataloged as Eris) was discovered 1-5-05 using data from 10-21-03 by Astronomers Michael Brown, Chad Trujillo and David Rabinowitz.**

The discovery of Xena (Eris) raised the urgency for a formal astronomical definition of the word "planet" because Xena was thought to be larger than Pluto. **This meant if Pluto was a planet, then so too was larger Xena (Eris).** Note: Eris was later calculated to be smaller but more massive than Pluto.

So what Exactly is a Planet (Greek origin for "wanderer")? Fundamentally, people understood that planets are spherical, can have moons, must orbit the Sun. But Pluto, Ceres, and Eris met these fundamental planetary **criteria**.

In August 2006, the 26th General Assembly of the IAU was held in Prague (capital and largest city of the Czech Republic), with 2500+ astronomers participating in symposia, joint debates/discussions, and special sessions. **A high-priority topic for debate & voting was formally defining the word "planet".**

Prior to the General Assembly, the IAU Executive Committee had formed a smaller 7-person Definition Committee tasked with formulating a definition & voting resolution for the IAU voting members. This committee convened in Paris late June into July 2006 to review and debate the years of prior efforts into planet definition.

On August 24, 2006---the last day of the 26th IAU General Assembly, **Six resolutions were voted upon** during the closing ceremony, including: **Resolution 5A: Definition of "planet" [within our Solar System; the vote was not counted but "passed with a great majority"]; and Resolution 6A: Definition of Pluto-class objects [vote passed with 237 votes in favor and 157 against and 17 abstentions].**

The International Astronomical Union's first official definition of the word "Planet" (in our Solar System) is defined as a celestial body that meets 3 criteria:
1. Is in orbit around the Sun
2. Has sufficient mass for its self-gravity to overcome rigid body forces [crush itself] so that it assumes a <u>**hydrostatic equilibrium**</u> (nearly round [spherical]) shape
3. Has cleared the neighborhood around its orbit

It was understood that #2 meant not so massive as to trigger nuclear fusion at its core thus becoming a star. **The IAU defined a new distinct class of objects: "Dwarf Planet".** These included non-satellite bodies meeting only the first two criteria, and they clarified that dwarf planets are not planets. **The first new members of the dwarf planet category were Ceres, Pluto and 2003 UB313 (later named Eris).** Two other KBOs were added later: <u>**Makemake**</u> and <u>**Haumea**</u>. The IAU maintains a "dwarf planet" watchlist which is ever-changing as astronomers discover new KBOs and these new candidates are evaluated.

The IAU recognized Pluto as "an important proto-type of a new class of trans-Neptunian objects." By this new definition, Pluto did not meet the 3rd requirement because it had not gravitationally cleared its orbit of bodies comparable in size other than its natural satellites. As of this writing, **there are 5 Solar System Objects classified Dwarf-Planets:** Ceres, Pluto, Eris, Makemake, and Haumea. **This will change in accordance with the Scientific Method** as new objects are discovered and evidence evaluated.
This is the essence of dynamic scientific discovery---exciting/open fair forum of discussion and debate!

Today's Solar System has 8 planets and 5 officially recognized "dwarf-planets" (Ceres, Pluto, **Makemake, Haumea**, Eris). **As of 2022, NASA.gov states that over 2000 KBOs have been cataloged.** (Credit: NASA).

Lesson 16: The process of scientific discovery is an unending, fluid flow of change and learning.

OK. Let's pause and take a closer look at a Very Important Universal Concept:

The Law of Cause and Effect,
which governs the world and the entire Universe!

The Law of Cause and Effect can be stated in many ways:

1. **Sir Isaac Newton called it his 3rd Law of Motion:** For every action, there is an equal and opposite reaction. And every effect has a specific cause, and every cause initiates a specific effect.

   A NASA and SpaceX example is a Rocket Launch (see left photo below) ---the rocket exhaust is blasted out of the bottom of the rocket with a force greater than the weight of the rocket, and the rocket accelerates in the opposite direction (up).

   Other examples: In boating, the paddle or propeller pushes the water backward and the boat moves forward. **In swimming,** if you push the water one direction, you will move in the opposite direction.

The 3 images below nicely illustrate the Law of Cause and Effect

  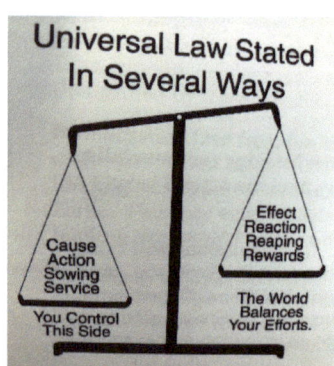

Rocket Launch (NASA)   **Corn Field** (www.Britannica.com)   **Apothecary Scale** (K. Caruso)

2. **In Farming: As you sow (seeds of corn for example) so will you reap (harvest of corn).** The land doesn't care what you plant, but it will return what you plant in the form of a harvest. This takes time through the seasons. And, sometimes storms and disasters wipe out that year's harvest. The land returns to the sower not exactly what is planted, but much more: an abundance, multiplied! This is how a few farms can feed millions!

3. **The <u>Apothecary Scale</u> is a balance scale. The bowl on the left is what you control: Causes, Action, Sowing, Service. The bowl on the right is what the world matches, i.e., the Effects, Reaction, Reaping, Rewards in response to your causes.** If you provide great service to many people, you will receive great rewards to match your service. The key is this: Focus on your service, which is a cause, and the rewards will be matched by the world as the natural effect or response to your service. Cause...Effect.

Action...Reaction. You don't control the rewards (the reaction), only your service (the action or cause). **Important thought about how much you are paid for any job:** Author and speaker Earl Nightingale said: **Your Income will be in exact proportion to** (a) the demand for what you do, (b) how well you do it, (c) the difficulty of replacing you.

4. **Regarding Attitude: What you send out (your attitude toward others) will determine what you get back (from others)---**if you are friendly and optimistic, this is what others will generally reflect back to you. If, however, you send out negativity, this is what you'll get back from others. **Only YOU Control Your Thoughts---that's all you need to succeed!**

   The key is this: You get to decide and choose your attitude from the moment you wake up, and the world will generally reflect back what you send out. We mentioned attitude earlier, and how important it is to choose a good one each day, especially if you want cooperation from others. How do people generally respond to you daily?

5. **In Business: The rewards or profits we receive will be determined by the level of service we provide and the number of people we serve.** This is why the company Amazon is doing so well financially---they serve millions of people 24/7.

6. **In Cooking, if you want a successful cake, you need to follow a specific recipe. The recipe doesn't care who's doing the baking,** your experience level, your background, your nationality, your age, etc.. **If you follow the recipe exactly, you will get the same excellent results: a nice cake!**

Question: What does this have to do with anything? Answer: This law affects your entire life! Understanding this Law is the Key to understanding how the world and science work, and the key to understanding that **things happen for a reason, and you can have much control over your own life...the universe operates on Law, and the Law is operating 24/7!**
You can become whoever and whatever you choose and want to become by taking action!

This Law provides us the Recipe we need to achieve anything we desire very much. If we'll just take the same actions or follow the same recipe others have followed to get results, we can also achieve those same results. **This also means if we keep doing what we've been doing, we'll keep getting the same results---no change. And if we want different results, we need to take different actions,** to change our future. **This is according to law, not luck, not chance.**

If you want a particular effect or result in your life, then all you need do is find who else has already achieved the same results you want, find out what steps they took (the recipe) to achieve those results, THEN, follow the same steps (recipe), remembering that it takes time---like the seasons. Now that you understand this important Law, let's continue our story.

> **Lesson 17: The Law of Cause and Effect governs how the universe works,** not just in physics and motion, but in planting, swimming, baking, farming, working, etc.. It is a key to understanding how the world works. Use this Law to achieve your goals.

Chapter 7.5: Science is about Disagreement!

> "You never change things by fighting the existing reality. To change something, build a new model that makes the existing model obsolete."
> – **Buckminster Fuller**

Science is about making careful observations, gathering data, seeing possible patterns, proposing ideas and hypotheses or predictions based upon the data, developing tests to validate the hypotheses, experimenting, failing, openly discussing, listening, disagreeing with other scientists, welcoming others to independently test and validate or discredit the hypotheses, modifying and refining the hypotheses to fit new data and evidence, continuously refining. **Everyone can participate and anyone can invalidate anyone else's hypotheses with new evidence and data. A single exception can make a hypothesis obsolete! It's completely transparent and fair!**

As new information is discovered, the best-fitting theories obsolete older theories. So, as more Kuiper Belt Objects beyond Neptune began to be discovered, more worlds of similar size & composition as Pluto, it became apparent that Pluto was no longer the 9th planet, but the first (closest) Kuiper Belt Object---the 3rd great class of solar system objects.

Thus, Pluto was reclassified. This is Exciting Living Science---the Scientific Method---in Action!

Let's look at more examples of concepts which did not exist or have drastically changed over time through scientific research, experimentation, better instruments, testing, discoveries: What else can you think of now or for the future (teleportation; warp speed; aliens; cloaking)?

9th Planet Pluto; The Earth does not move
Tele/Microscopes; Earth is flat, is a planet
Recorded Sound/Voice; Telephone
Electricity; Batteries; Electric Vehicles
Gravity; Big Bang; Quantum Mechanics
Submarines; Aircraft; Drones; Spacecraft
Periodic Table of Elements; Fission; Fusion
Evolution; Fossils; Dinosaurs; **SCUBA**
Other **Galaxies**; Other Solar Systems
Radio; Television; Cable; Artificial Satellites
Asteroids/Asteroid Belt; Dwarf-Planets
Microbes; Medicine; Viruses; Covid-19
The Kuiper Belt; Oort Cloud; Spectrum
Red Shift; Infrared Light; UV Light
Astronauts; Space Stations; Living in Space
Electromagnetic Spectrum – **ROYGBIV** occupies a tiny portion of the entire (mostly invisible) spectrum
The people of ancient times only knew about 5 visible planets: Mercury, Venus, Mars, Jupiter, Saturn.

The Earth is the center of the universe; scale models; globe
The Sun and all of the stars and planets orbit the Earth
Continental Drift; Plate Tectonics; Earth's Interior Structure
Planets other than Mercury, Venus, Mars, Jupiter, Saturn
LASERs; LEDs; Magnetism; computers; **Relativity; E=mc^2**
Robotics; Artificial Intelligence; Mars Rovers; **SkyCrane**
Automobiles; Autonomous Vehicles; Global Navigation (GPS)
Milky Way Galaxy; Interplanetary Space Probes; tablets
Extra Solar Planets (orbiting other suns); Cell Phones; IoT
Trans-Neptunian Objects/**Kuiper-Belt Objects**; **3D printing**
Black Holes/Supermassive Black Holes; Pulsars; **the Internet!**
Atoms/Subatomic Particles/Nuclear Power; Carbon-14 Dating
DNA; Genetics; Blood Cells; Plant Cells; Cloning; Bar Codes
X-Rays; Radioactivity; Nuclear Power; Atomic Clocks; CATscan
Transistors; Electronics; Superconductors; Supercomputers

The Scientific Process is often referred to as the Scientific Method, a method or open forum of sharing ideas (theories or hypotheses), asking questions, which are open and subject to independent challenging, testing hypotheses, observation and verification by anyone.

Here's a Diagram of The Scientific Method:
Note: Pluto's re-classification from Planet to Dwarf-Planet was the Scientific Method in Action!

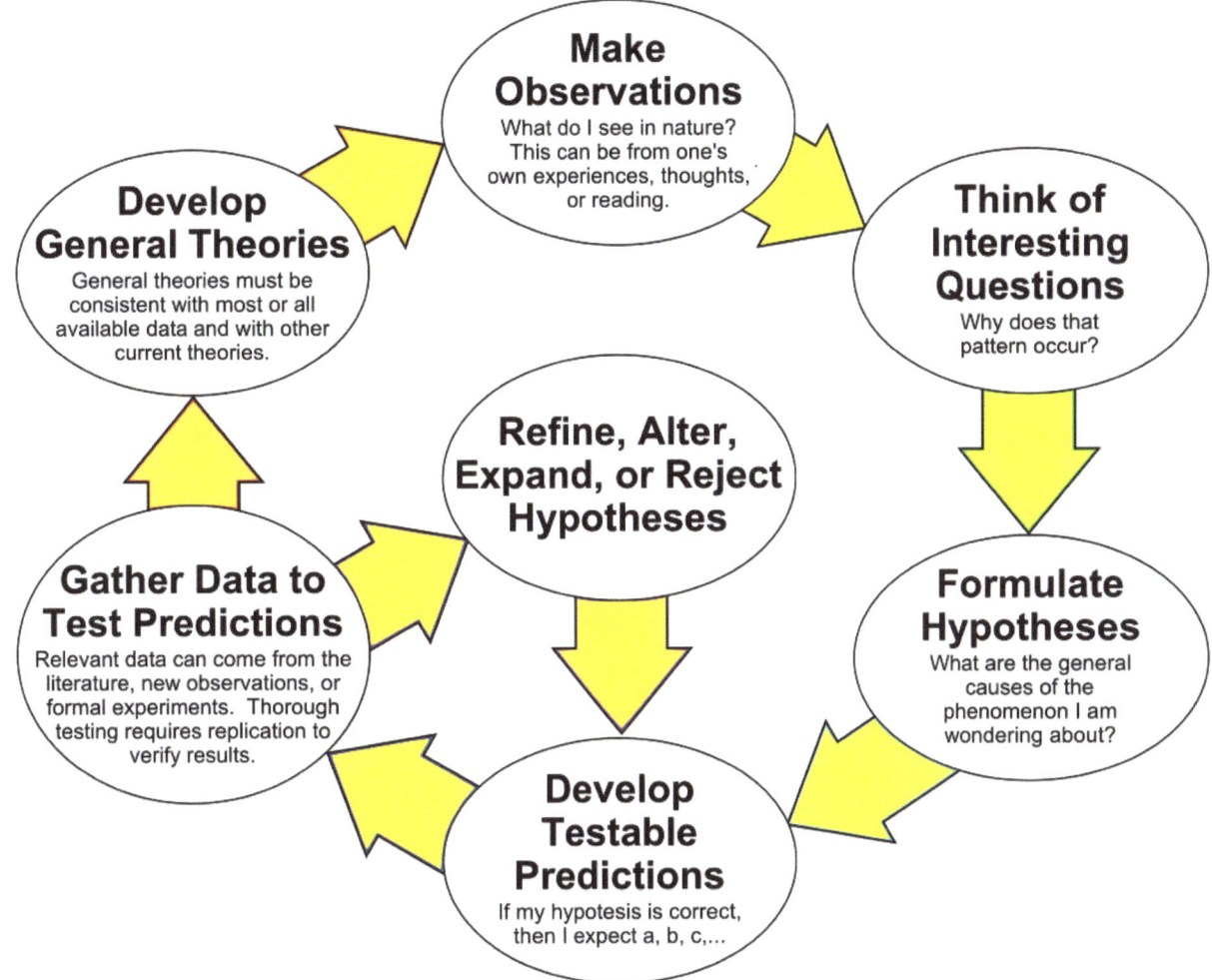

(Credit/Source of this diagram of the Scientific Method):
https://upload.wikimedia.org/wikipedia/commons/thumb/5/5c/The_Scientific_Method_as_an_Ongoing_Process.svg/1200px-The_Scientific_Method_as_an_Ongoing_Process.svg.png

> "Science…is made up of mistakes, but they are mistakes which it is useful to make, because they lead little by little to the truth." – Jules Verne from "A Journey to the Center of the Earth"

Let's pause for an Amazing True Story about the Scientific Method, Experimentation, "Failing" over 10,000 times, and Persisting Until Success!

Would you agree that Thomas Edison (1847-1931) was an amazing inventor? He was awarded 1093 U.S. patents and 2332 global patents in all, in the fields of electric light and power, phonographs and recorded sound, telegraphy and telephony, batteries, mining and iron ore milling, cement, motion pictures, and more. **Did you also know that Thomas Edison "failed" more than most people ever try?**

Author Napoleon Hill interviewed Edison about his 10,000+ light bulb experiments. In developing his incandescent light, Edison maintained 2 stacks of logbooks detailing his **10,000+ experiments, each of which failed:** a glass bulb containing a wire through which electricity flowed, creating a hot glowing filament---**his filaments kept burning up!** After exhaustive experimenting one day, Edison awoke from a nap with the solution: He combined the known concepts of a glowing filament AND the glowing charcoal embers of a dying campfire. He pumped the oxygen from the bulb & the wire no longer burned (due to lack of oxygen)! **Edison considered his "failed" experiments Learning Experiences toward a solution!**

In his audio seminar *"Success Mastery Academy"*, author Brian Tracy shares this Edison fable:

> While in pursuit of a successful incandescent light, Edison performed over 10,000 experiments which did not work. **(That part is true.) The Fable about Edison's Attitude continued:**
>
> After Edison's 5000th experiment, a reporter visited and asked, "Mr. Edison, why do you persist with these foolish experiments? You have failed 5000 times!"
>
> **Edison replied, "Young man, you don't understand how the world works. But I want to tell you something. I have not failed at all! I have successfully identified 5000 ways which do not work, which puts me 5000 ways closer to the way that will work. How many do you know?"**
> (See Napoleon Hill Foundation: https://www.naphill.org/tftd/thought-for-the-day-04-04-20/)

What a great attitude and philosophy! If you learn from your failures, they help you toward your goals! They count! Ask questions. This is a great way to approach obstacles which you'll face regularly!

The Menlo Park Museum shares this story:
(**https://www.menloparkmuseum.org/history**)
Menlo Park NJ is where Edison founded his Research and Development Laboratory:

Thomas was a very curious child. There's that trait (Curiosity) again! Thomas liked to understand how things worked and he asked many questions, so many questions that some of his early teachers became annoyed with his questions.

He didn't do well in formal school, so his mother decided to home school him starting at the age of 12. **She noticed and nurtured his curiosity and interests in chemistry and electronics** and provided Thomas his 1st scientific book: "The School of Natural Philosophy", with which he performed every experiment listed in the book.

The Franklin Institute of Philadelphia says that Edison credited his mother as a great teacher who encouraged him to learn and self-study.
(**https://www.fi.edu/history-resources/edisons-lightbulb**)
(Credit: The Franklin Institute of Philadelphia)

Thomas Edison in his Menlo Park NJ Factory in 1879, holding his incandescent light bulb.

Interesting Fact: Thomas Edison was a Self-Taught Subject-Matter-Expert---as was Percival Lowell, Clyde Tombaugh, Jacques Cousteau, the Wright Brothers & Napoleon Hill, through their actions!

Chapter 8: A Heart Bigger Than Texas!

"Somewhere, something incredible is waiting to be known." – Carl Sagan

Navigational Scientists are Mathematical Geniuses! At launch in 2006, **New Horizons was "aimed" at an empty place in space where Jupiter would be on February 28, 2007** (for a gravity assist to increase its speed to about 51,000 MPH (83 Km/hour) and thereby reduce its travel time to Pluto by 3 years!

Then they navigated New Horizons toward an empty place in space where Pluto would be on July 14, 2015---8 years in the future, so they would both arrive at the same place at the same time! WOW! **Now THAT'S IMPRESSIVE MATHEMATICS and PRECISION NAVIGATING!** All thanks to the Natural Laws of Physics which operate the same way on Earth as they do throughout the Universe!

Here's the **"fail-safe"** image which New Horizons captured on July 13, 2015 a couple of days prior to its closest approach. Fail-safe means this was a **contingency** photo taken just in case New Horizons collided with and was destroyed by unknown space debris in the Pluto System during closest approach---New Horizons was covered in aluminum shielding and bullet-proof Kevlar wrap. Team leader Alan Stern wanted to be certain to capture at least one close-up image of Pluto which would be transmitted immediately to Earth before closest approach. After a 9.5-year journey, the team wanted at least 1 good picture just in case everything went wrong during the closest approach through the Pluto system!

Here's what New Horizons captured and transmitted to Earth. PLUTO HAS A HEART BIGGER THAN TEXAS! Well, it's really an enormous glacier & icy plain shaped like a heart, with mountains.

"**Fail-Safe**" photo of "Dwarf-Planet" Pluto by New Horizons as it approached on July 13, 2015.
(Credit: NASA/Johns Hopkins University Applied Physics Laboratory/Southwest Research Institute)
Watch Time-Lapse Video as New Horizons' Approaches Pluto: https://www.youtube.com/watch?v=xmqDpuDLVYw

> **Lesson 18: Prepare well for success. But plan for the worst & for contingencies.** Make sure you have emergency plans A, B, C and D.

July 2015: Clyde Tombaugh's children **Alden Tombaugh** (center) and **Annette Tombaugh** (right) pose with a photo of "dwarf-planet" Pluto with **Sylvia Kuiper** (left), daughter of **Gerard Kuiper**, the astronomer for whom the Kuiper Belt beyond Neptune is named. (Credits: NASA, JHUAPL/SwRI, and "Clyde Tombaugh, Discoverer of the Planet Pluto – Academy of Achievement," 2019, https://achievement.org/achiever/clyde-tombaugh/ , page 1).

Because New Horizons' radio signals from the distance of Pluto would take about 4.5 hours to reach Earth (1-way), the entire reconnaissance of the Pluto System would have to be pre-programmed, not live. This included all camera angles, target photography, science experiments! That's precision planning! Again, this is incredible talent…New Horizons was moving 30 times faster than a bullet!

So, the team practiced taking images and collecting science data years before, starting with the flyby/gravity assist at planet Jupiter on February 28, 2007---8 plus years before the Pluto encounter in July 2015! Jupiter increased New Horizon's speed by 4 kilometers/second or about 9000 miles per hour, to about 83,000 km/hr (about 51,000 mph), shortening New Horizon's trip to Pluto by about 3 years!

> **Lesson 18.4: Leverage the assistance of others** to help you take years off your journey toward your goals.

The New Horizons Team had rehearsed and tested their equipment, cameras, programming, camera angles, and targets, 8+ YEARS in advance---and full "dress rehearsals" in 2012, 2013, 2014, 2015 months, weeks, and days before the Pluto encounter. It's a good thing they did, because they encountered several safe-mode shutdowns and a scary 90-minutes of silence during which New Horizons main computer failed and switched over to the backup computer because of Jupiter's intense radiation, which crippled the main computer.

> **Lesson 18.8: Prepare, Practice, Rehearse for important events or long journeys.** Develop back-up or contingency plans in case something goes wrong (like New Horizons' 2nd computer)! Develop emergency plans A, B, and C---but at least A. Think about the Apollo Astronauts who landed on the Moon from 1969-1972---they practiced and rehearsed every part of the mission many times. **How do students prepare for musicals, plays, or sports games?** They practice and rehearse well before the event. That helps determine success!

Think of the Engineering and precision planning and programming which went into the automated flyby rehearsed years, months, weeks, and even days before the Pluto close encounter! This was 9.5 years in the making! The graphic below shows Pluto's pre-programmed flight path through the Pluto System, a carefully orchestrated sequence of events which had to execute flawlessly to gather the most images and scientific data. **This is brilliant navigation, computer programming, and teamwork at its best!**

(Credit: NASA)

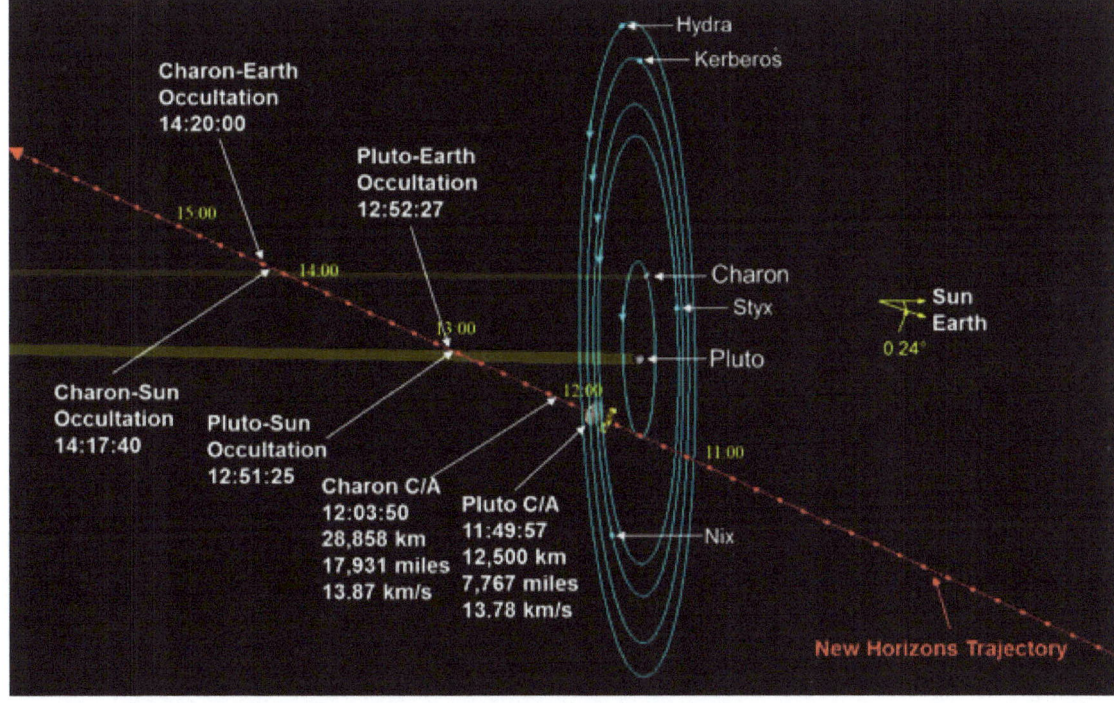

New Horizons, traveling about 51,000 miles+ per hour (83,000 KPH) had to be pre-programmed by brilliant software programmers working closely with astronomers and scientists to be able to photograph Pluto and its 5 moons, during a fast flyby and before it flew too far from Pluto. Remember, New Horizons was not stopping to go into orbit around Pluto. **New Horizons was flying past Pluto at about 30 times faster than a bullet (based upon 1700 mph for a bullet)!**

All the digital photos and science data which New Horizons gathered during its several days of encounter with the Pluto System were stored in its onboard memory for later transmission to Earth. **Transmitting all this information back to Earth took nearly 18 months after the flyby!**

Mission Accomplished! Pluto Explored!
And Improved Pluto Postage at Last!

On May 31, 2016, after much of the Pluto exploration data had been transmitted back to Earth, the United States Postal Service issued this new block of stamps: **PLUTO EXPLORED!** to Celebrate the accomplishments of Pluto's Exploration by the New Horizons spacecraft, OBSOLETING the former 29-cent Postage Stamp stating: "Pluto...Not Yet Explored". (Credit: United States Postal Service (USPS)).

Note: The process of the New Stamp obsoleting & replacing the former 29-Cent Stamp based upon new data & observations is very much like the process of The Scientific Method, in which old theories are replaced by theories which better-fit the data and observations.

Lesson 18.9:
Celebrate your Successes!

New Horizons' approach photo of Pluto and Charon in a single picture frame showing relative distance and scale. (Credit: NASA, JHUAPL, SwRI)

And remember Jim Christy's grainy bump photo of Charon from 40 years earlier (page 26)? Here's what Charon looks like! **Jim, his wife Char and the world were excited to see Charon up close at last!** The large chasm across the width of Charon has walls about 4 miles high! Scientists theorize this may be due to a former ocean just beneath the surface which evaporated and collapsed, leaving high cliffs.

Lesson 19: If you seek, you will find. The more you search, the more you'll find (and learn and discover).

Dr. Brian May at Pluto Event (Credit: NASA/JHUAPL/ Edward Whitman/SwRI)

Google: Charon's red polar cap of **tholins.** (Credit: NASA/JHUAPL/SwRI)

Pluto's 5 moons to scale. (Credit: NASA, JHUAPL, SwRI)

Nix and Hydra
Were discovered in 2005, Styx (goddess of the underworld river) in 2011, and Kerberos (3-headed dog guarding Pluto's realm) in 2012.

The **Rock Band Styx** appreciated the internet-crowd-sourced moon naming and posed in a photo with discoverer Mark Showalter. **Queen** lead guitarist & **Astrophysicist Dr. Brian May** also participated in the Pluto celebrations on July 18, 2015!

Chapter 9: New Horizons, Rocking the Solar System

"I think it's exciting that all the textbooks will have to be rewritten."– Alan Stern

Close up view of **Wright Mons, one of two potential cryovolcanoes spotted on the surface of Pluto** by the passing New Horizons spacecraft in July 2015. Close-up images of a region near Pluto's equator reveal a range of youthful mountains rising as high as 11,000 feet (3500 meters) above the surface. Age estimate is 100 million years because of lack of craters. *(Credits: NASA/JHUAPL/SwRI/Jeff Moore)*

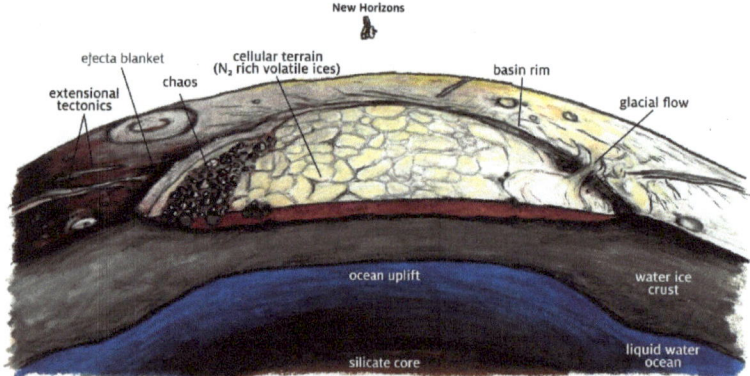

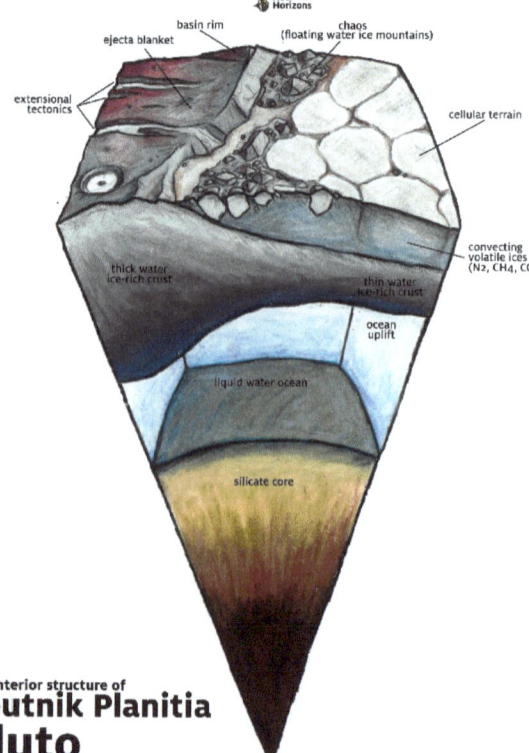

https://www.nasa.gov/sites/default/files/thumbnails/image/top10_2sputnikart.jpg
Illustration of Sputnik Planitia at Pluto (above) and Illustration of Interior Structure of Sputnik Planitia at Pluto (right).
(STEAM Artist Credits: James Tuttle Keane (NASA JPL Planetary Scientist & Artist on New Horizons Team)

See "10 Cool Things we learned about Pluto dated July 14, 2020 at this NASA site:
https://www.nasa.gov/feature/five-years-after-new-horizons-historic-flyby-here-are-10-cool-things-we-learned-about-plut-0
Also see New Horizons' Approach to Pluto at https://www.youtube.com/watch?v=xmqDpuDLVYw

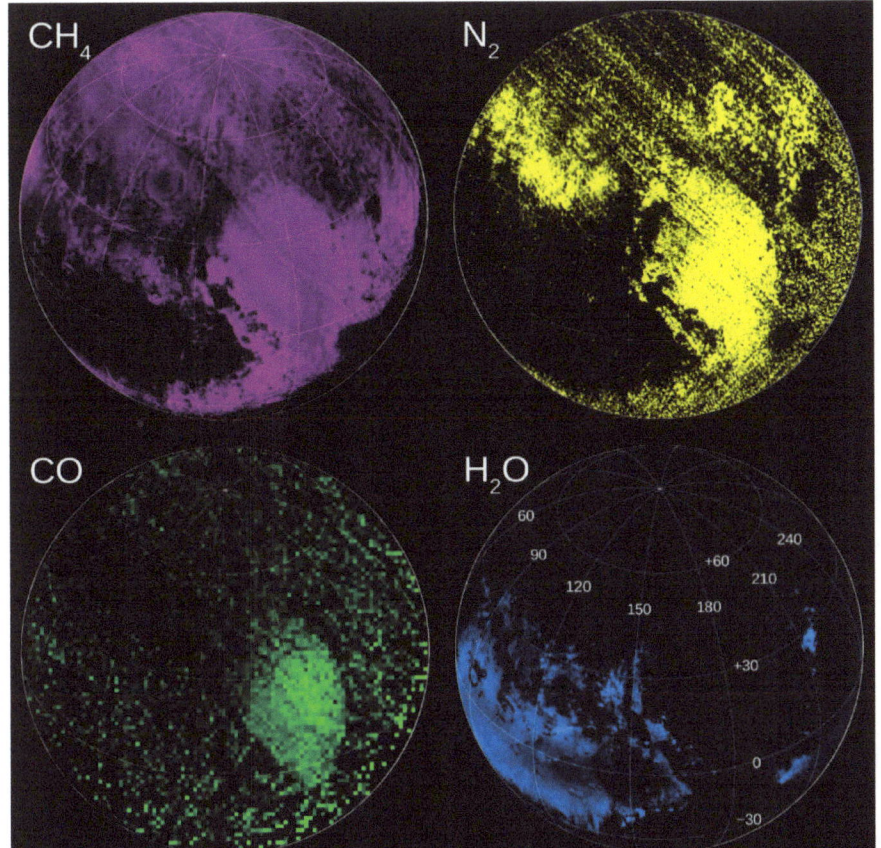

The powerful instruments on New Horizons not only gave scientists insight on what Pluto looked like, their data also confirmed (or, in many cases, dispelled) their ideas of what Pluto was made of. These compositional maps – assembled using data from the Linear Etalon Imaging Spectral Array (LEISA) component of the Ralph instrument – indicate the regions rich in ices of methane (CH_4), nitrogen (N_2) and carbon monoxide (CO), and, of course, water ice (H_2O). *(Credits: NASA/JHUAPL/SwRI and Ames Research Center/Bill Keeter)*

Below: Snip from "Pluto Flyover Movie", a Wow! artistic 3D rendering highlighting terrain based upon data from MVIC and LORRI instruments aboard New Horizons. *(Credits: NASA, JHUAPL/SwRI, Lunar and Planetary Institute in Houston, Digital Mapping & Rendering by Paul Schenk and John Blackwell of Lunar and Planetary Institute.)* **See You Tube Pluto Flyover** *https://youtu.be/g1fPhhTT2Oo*

In his audio learning seminar, "Success Mastery Academy", author and motivational and self-development public speaker Brian Tracy shared a Fun Fable about Famous Physicist Albert Einstein when he occupied an office at Princeton University's mathematics building in the 1930s:

The Story Goes…Albert Einstein once gave an exam to an advanced class of physics students. After the exam, while walking back to his office, the teaching assistant who had helped administer the exam asked Einstein, **"Dr. Einstein, wasn't that the same exam you gave to the class last year?"** Dr. Einstein replied, "Yes it was the same exam." The assistant then asked, "But Dr. Einstein, how could you give the same exam to the class two years in a row?" **Einstein smiled in reply, "Because in the last year, the answers have changed."** Key Story Point: Answers and Knowledge are changing rapidly in most fields!

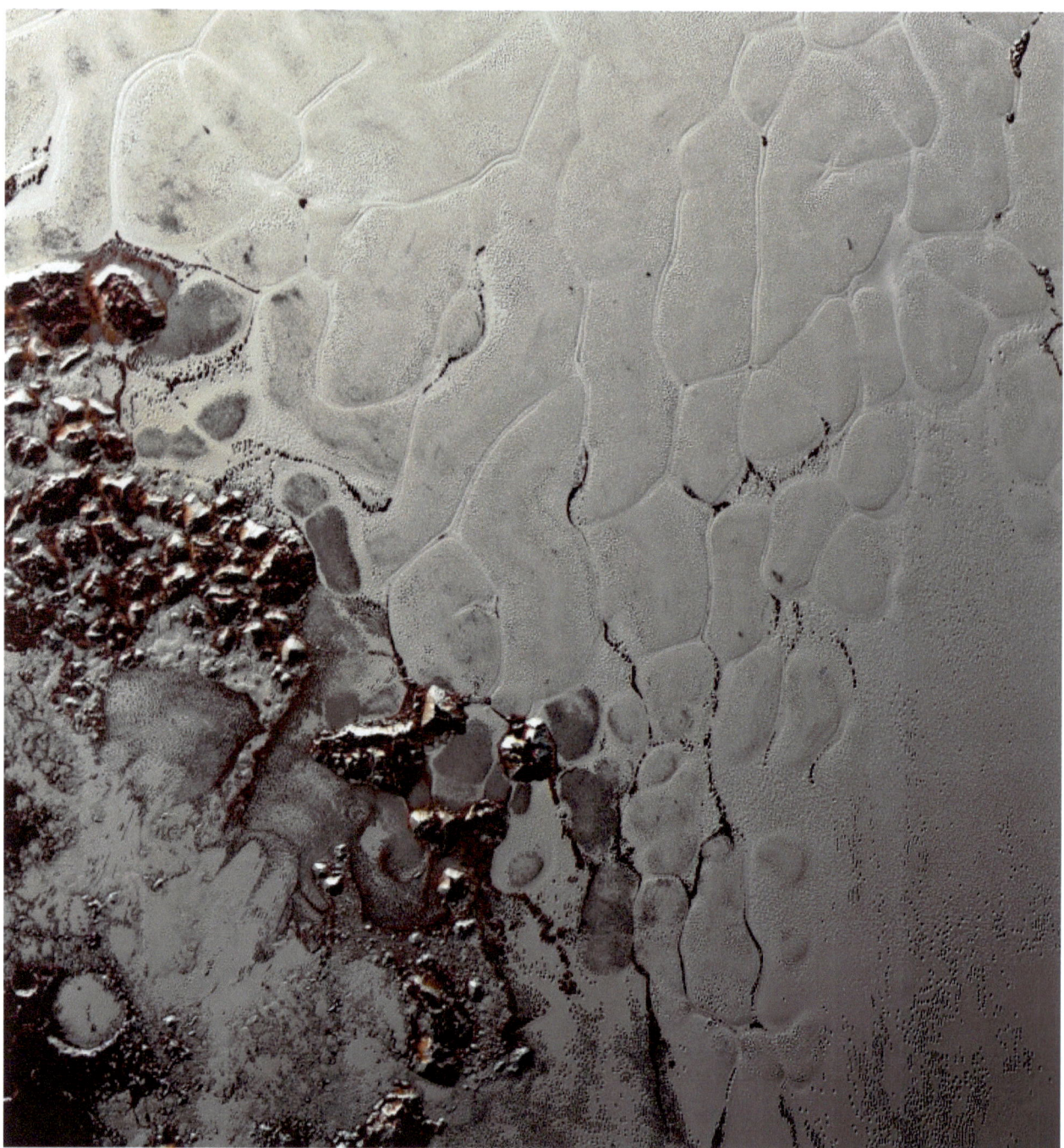

Scientists from NASA's New Horizons mission used state-of-the-art computer simulations to show that the surface of Pluto's Sputnik Planitia is covered with churning ice "cells" that are geologically young and turning over due to a process called convection. The Nitrogen ice flows around & between the solid water-ice mountains. (*Credits: NASA/JHUAPL/SwRI*)

> **Lesson 19.1: Ignore Limits!** <u>Roger Bannister</u> ran the sub-4-minute mile when others said it was impossible.
>
> > His record lasted just 46 days before being broken again---now that people knew it could be done!
>
> - The first two to climb to the top of Mount Everest, highest point above sea level on Earth were **Edmund Hillary** and **Tenzing Norgay** (a bee keeper in New Zealand).
> - The Wright Brothers made history on December 17, 1903 with their first powered flight.
>
> **Ignore Limits! They can be and will be broken!**

This detailed image of Pluto's Southwestern Sputnik Planitia shows thousands of pits in its surface of nitrogen ice and a cryovolcano at center bottom of the image. (*Credits for this and inset image: NASA/JHUAPL/SwRI*)

> **"Nothing in life is to be feared, it is only to be understood. Now is the time to understand more, so we may fear less."** – <u>Marie Curie</u> (Scientist who discovered radioactivity, and, together with her husband Pierre, discovered the radioactive elements polonium and radium. Also championed the development of X-Rays after Pierre's death)
>
> **"If I have seen further it is by standing on the shoulders of Giants."**
> – Sir <u>Isaac Newton</u> (1643 – 1727), English mathematician, physicist, astronomer, theologian, author. Established classical mechanics, contributed to optics, shares credit with <u>Gottfried Wilhelm Leibniz</u> for establishing a new field of mathematics called <u>Calculus</u>, inventing/developing calculus, invented the <u>reflecting telescope</u>, discoverer of the laws of Gravity and Motion and the equation for gravity, from <u>**"The Correspondence of Isaac Newton"**</u>

False-Color High Resolution Image of Pluto's Heart to enhance features captured by RALPH---amazing photography at a flyby speed of 30x faster than a bullet! (*Credits: NASA/JHUAPL/SwRI/ZLDOYLE*) Orange rectangle see p. 54 top.

> **Lesson 20:** "Your assumptions are your windows on the world. Scrub them off every once in a while, or the light won't come in." – Isaac Asimov
>
> "Any sufficiently advanced technology is indistinguishable from magic." – Arthur C. Clarke from the book "Profiles of the Future: An Inquiry Into the Limits of the Possible"
>
> **Lesson 21:** In his book, "A Sense of Urgency", author John Kotter recommends these four actions:
> 1. Accomplish something important every day.
> 2. Act with a sense of urgency in all you do.
> 3. Innovate. Look for opportunities, then take daily action on those opportunities.
> 4. Continuously improve.

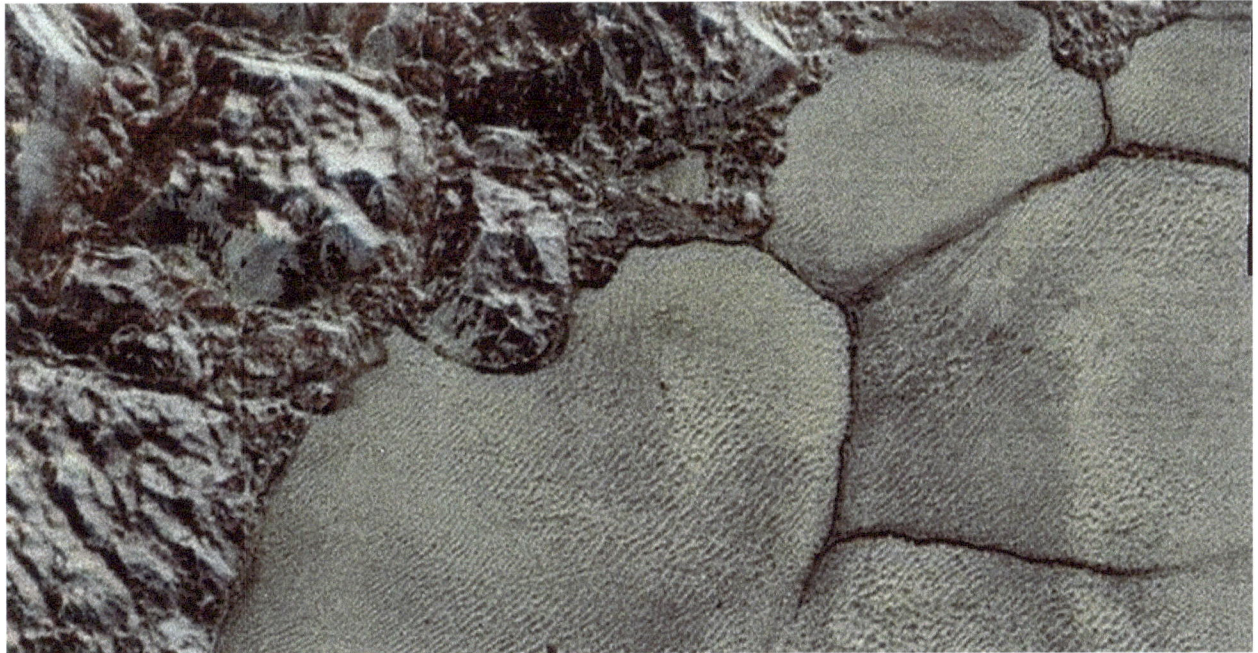

Extreme Close-up from previous page orange rectangle of water-ice mountains on the northwest fringes of Pluto's Sputnik glacier may provide the particles, and Pluto's beating nitrogen "heart" provides winds. *(Credits: NASA/JHUAPL/SwRI)*

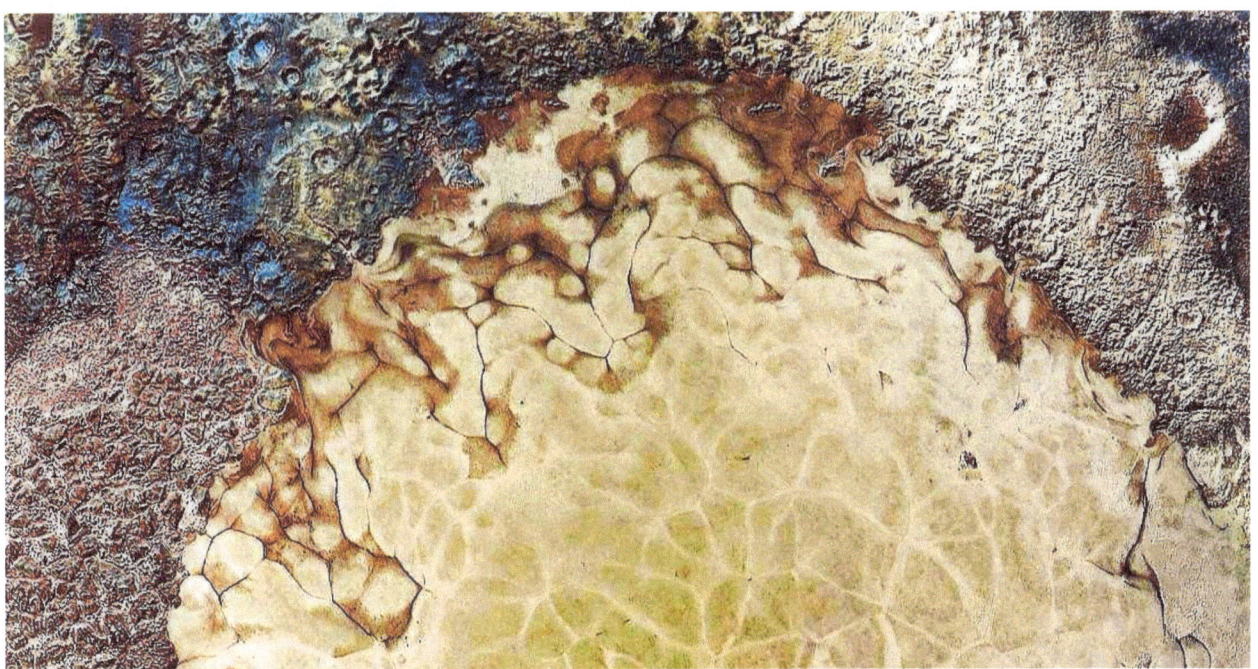

High-resolution False-Color view of Northern Sputnik Planitia glacier showing flow patterns, captured by NASA's New Horizons spacecraft. *(Credits: NASA/JHUAPL/SWRI)* See https://www.youtube.com/watch?v=JqmIo-tUd48 !

> "We are a way for the cosmos to know itself."
> – Carl Sagan, from **"Cosmos"**

After flying by Pluto and into its shadow, New Horizons turned its camera and science instruments around to view Pluto's atmosphere---turns out that Pluto's sky is Blue, like that of the Earth!
(Credits: above and below NASA/JHUAPL/SWRI)   **Below: Close-up of Pluto's Layered Atmosphere**

Lesson 22: We have more in common with others than we may at first realize! Next time, Start looking for what we share in common. Humanity, for example!

"Always use the word 'impossible' with the greatest caution." – Wernher von Braun
"The difficult can be done immediately. The impossible takes a little longer." – U.S. Army Corps of Engineers

Chapter 10: Honoring Pioneers!

"Dare Mighty Things!"
– NASA Jet Propulsion Laboratory (Mars Exploration Team, the **SkyCrane Maneuver**)

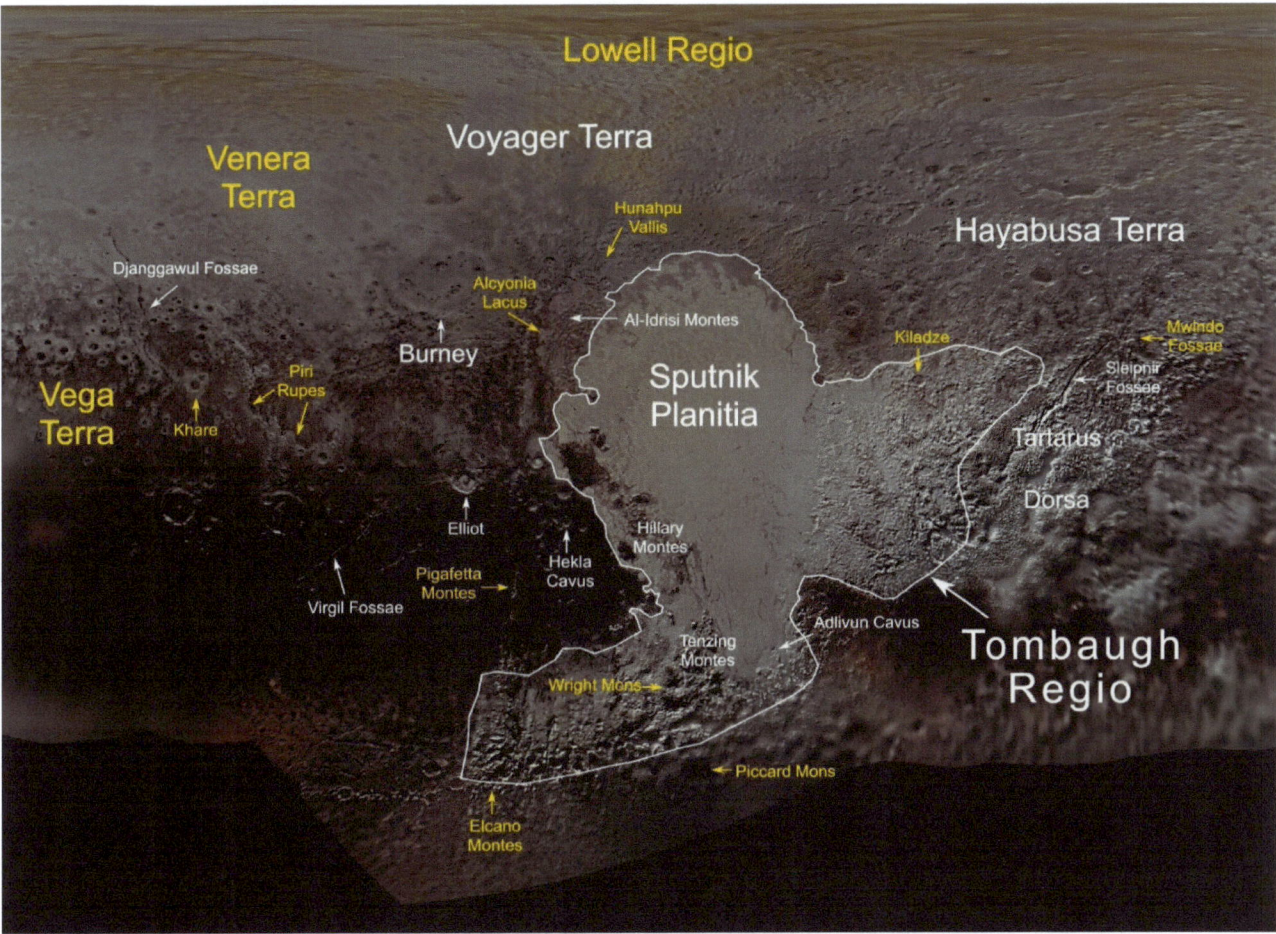

Pluto feature names officially approved by the **IAU** on Aug. 8, 2017 (Credit: NASA/JHUAPL/SwRI, IAU)

- **Pluto's Giant Heart is named "Tombaugh Regio" in Clyde's honor as Pluto's Discoverer.**
- **A large Crater is named in honor of Venetia Burney, who gave Pluto its name.**
- **Lowell Regio honors Percival Lowell.**
- **Wright Mons honors the Wright Brothers.**
- **Sputnik Planitia honors the first satellite (Soviet/Russian) to orbit the Earth.**
- **Venera Terra honors the first unmanned spacecraft (Soviet/Russian) to land on Venus.**
- Edgeworth crater honors the Irish astronomer who proposed the disc of bodies beyond Neptune, later confirmed by Gerard Kuiper in 1992, now called the **Edgeworth-Kuiper Belt**.
- Hillary Montes and Tenzing Montes honor Sir Edmund Hillary of New Zealand and Tenzing Norgay a **Sherpa** Tibetan guide from Nepal, who became first to conquer Mount Everest (8888 meters high) on May 29, 1953.
- Voyager Terra in honor of **Voyagers 1 & 2** spacecraft, which explored the outer solar system.

**You can look up the additional names on Google, Bing or Wikipedia by searching:
Names of Features on Pluto and its Moon Charon!**

As technology & telescopes improved, so did our view of Pluto, as shown below, until New Horizons finally visited up-close. From a tiny spec of light in 1930 to a world with a heart in 2015!

(Credits: Lowell Observatory Archives/Hubble Space Telescope Science Institute/NASA/JHUAPL, SwRI)

High-resolution False-Color view of Eastern Sputnik Planitia glacier showing flow patterns and rough Eastern terrain, captured by NASA's New Horizons spacecraft. *(Credits: NASA/JHUAPL/SWRI)*

> **Lesson 23: Improve your vision! Refine your philosophy as you learn more and get closer to achieving your goals.**
>
> If necessary, get a better instrument. Get closer! Enlist the help of others---leverage their strengths and experience. Remain flexible, agile, open to change. Have crystal clarity of purpose!

The Power of Imagination + Clarity + Action!

Walt Disney World's Spaceship Earth at EPCOT Center, Orlando FL. (Credit: Walt Disney World)

Walt Disney World Resort opened its **Experimental Prototype Community of Tomorrow (EPCOT)** on October 1, 1982. Many years before, Walt Disney had imagined EPCOT and started making detailed plans to build it. But he died on December 15, 1966.

At EPCOT's opening, Walt's Nephew **Roy E. Disney** was interviewed by a reporter. The story is told that the reporter said, "You know Roy, it's a shame Walt never got to see EPCOT."

Roy replied, "Walt saw it first, that's why you're now seeing it."

Where did Walt Disney see it first? In his Imagination!

Roy was expressing a concept which many artists, athletes, professionals of all types, parents, teachers, musicians and other creative individuals understand, but which many people on planet Earth never realize:

> **Everything made by human beings is created twice: First in the mind (the imagination) of an individual as a thought, and then in physical reality (through work and action).**

Examples include: Your clothes, your phone, the light fixtures and furniture in your house, the house you live in, the car you drive, the businesses all over the world---**all began as a thought or idea in 1 person's mind, and then that person put their energies and efforts into that vision and, with the help of others, brought that idea into reality through work.** A thought is essentially a 3-dimensional preview in the mind of a single person---you might say it's like a movie theater "Coming Attraction".

This book began as a thought in the author's mind, then as notes on paper, then photos and ideas gathered in a book, and finally typed chapters on a laptop computer. This is how the human mind works, and it works for any idea we want with a burning desire to bring about if followed by action. **And you "own" such a mind!**

Composer, musician, and recording artist Yanni, described this "vision process" during a live performance in September 1993 at the 2000-year-old Herodes Atticus Theater at the Acropolis of Athens Greece:

> "Everything great that has ever happened to man since the beginning, has begun as a single thought in someone's mind. And if any one of us is capable of such great thought, then all of us have the same capacity and capability because we're all the same."

And you can do the same! That's how people turn "nothing" (i.e., thoughts) into "something" (i.e. things), because Thoughts are Causes and Conditions are Effects! Action, Reaction!

Let's talk about Success Philosophy!

Andrew Carnegie (Carnegie Steel industrialist who sold to U.S. Steel in 1901, philanthropist, and wealthiest man in the world in 1908) described a success philosophy to 25-year-old Napoleon Hill who had arranged to interview Carnegie for a magazine article about successful business leaders. After 3 hours of interviewing, Carnegie invited Hill to his mansion to continue the interview for 3 days!

Like the story of the Lowell Observatory when Clyde mailed his sketches: Unknown to Napoleon at the time, **Carnegie had been seeking an author willing to devote his life** to gather the **"World's 1ˢᵗ Practical Philosophy of Individual Achievement" based upon the combined wisdom of 504 of the most accomplished Americans alive at the time,** then publish the results and share it with the world so others could use it for their own benefit. Carnegie told Hill it would be a tragedy for the success principles of so many accomplished individuals to go to the grave when these people died.

At the end of the 3-day interview, Carnegie asked Napoleon if he'd be willing to devote the next 20 years---because that's how long it would take**---to interview & learn the success wisdom of these accomplished Americans, with letters of introduction to be provided by Carnegie, with travel expenses paid, but no more, and to organize this philosophy and share it with the people of the world. Yes or No? Carnegie was hiding a pocket watch and had secretly given Hill just 60 seconds to respond. Carnegie later shared with Hill that successful people make decisions quickly and change them slowly.**

How would you have responded? (Remember Lessons 4 & 6 Be willing to Act without Credentials, and when an amazing opportunity arises, Take It, and the Corridor Principle will kick in). **Hill said his first thought was, "I couldn't do that! I don't have enough money or schooling." Then he thought carefully about who was asking him and what Carnegie saw in Hill that Hill didn't see.** After just 29 seconds, **Hill said, "Mr. Carnegie, I not only will accept your commission, but you may depend upon it, Sir, that I will complete it!" He earned the job! That decision improved the lives of millions of future readers!**

Who did Hill interview? Alexander Graham Bell, George Eastman, Harvey Firestone, Henry Ford, Thomas Edison, Charles Schwab, F.W. Woolworth, William Wrigley, Wilbur Wright, former President Woodrow Wilson, former President Theodore Roosevelt, former President William Howard Taft, Clarence Darrow, John D. Rockefeller, John Wanamaker, and more. **Hill published his research in 1928 (20 years after starting) as an 8-Volume "Law of Success", then as the 17 Success Principles in his 1937 best-selling book "Think and Grow Rich", with additional books too.**

Chapter 11: Next Goal? <u>Arrokoth</u> and Beyond!

"Go as far as you can see; when you get there, you'll be able to see farther."
— <u>J.P. Morgan</u>

What was next for the New Horizons spacecraft? Continue exploring of course! NASA approved funding to extend the mission to another KBO! In 2011, while still enroute to Pluto, the New Horizons Team asked for assistance from the Hubble Space Telescope scientists to help identify possible Kuiper Belt Objects (KBOs) to visit after Pluto which were within the fuel reserves of New Horizons.

New Horizons Team member **Marc Buie** discovered this KBO on June 26, 2014 using the Hubble Space Telescope. It was first nicknamed <u>Ultima Thule</u> – a mythical island beyond the borders of the known world, later named <u>Arrokoth</u> – a Native-American term meaning "sky" in the <u>Powhatan</u> language.

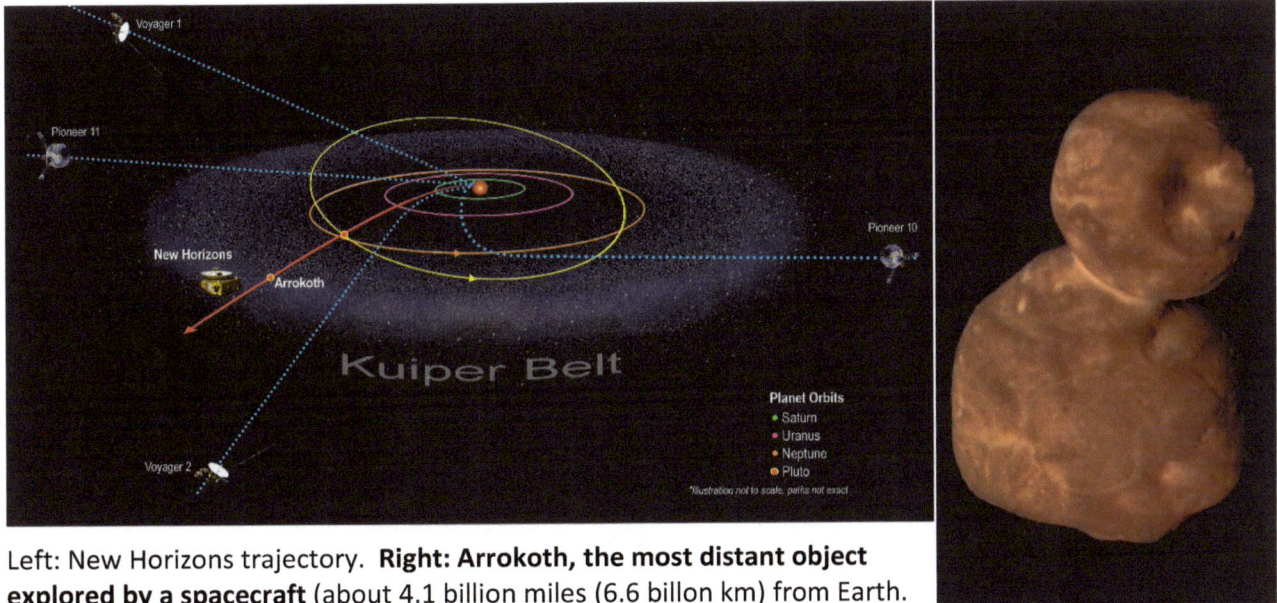

Left: New Horizons trajectory. **Right: Arrokoth, the most distant object explored by a spacecraft** (about 4.1 billion miles (6.6 billon km) from Earth. (Credits: NASA/ASA/JHAPL, SwRI, and right adding Thomas Appere) Arrokoth's Red Color? More tholins!

On January 1, 2019, about 3.5 years and 1 billion miles (1.6 billion kilometers) past Pluto, New Horizons spacecraft was navigated to Kuiper Belt Object "Arrokoth"---thought to be a remnant of the early solar system. Here's what New Horizons photographed from about 2198 miles (3538 km) away---two objects made of ice, dust and rock which gently collided and stuck together. It measures about 22 miles (35 km) long, 12 miles (20 km) wide and 6 miles (10 km) thick. Arrokoth's color, like Charon (p.48) is due to tholins (methanol (CH_3OH) broken down by sunlight/cosmic rays). New Horizons spacecraft was in good condition and still had fuel for exploring. The team proposed possibly performing heliospheric, astrophysics or planetary science from its unique vantage point within the Kuiper Belt.

> **Lesson 24: Even once you've achieved your primary goals, make sure you develop additional goals.** Stay curious and continue to learn and explore!
>
> After the first Lunar Landing of Apollo 11, **NASA quickly realized it needed to prepare future Moonwalking Astronauts with setting goals beyond their Moon landing missions.** The same will be true for the Return to the Moon and future missions to Mars.

Did you know there's a process of Goal-Setting?

Author and Success Speaker Brian Tracy and Author and Motivational Speaker Zig Ziglar provide some insights into setting goals: Brian says **Clear, Measurable, Time-Bounded Goals enable you to control the direction of change in your life.** He says only 3% of adults have written goals. In his book, "See You at the Top", Zig Ziglar (1926-2012) outlined these Steps:

Step 1: DECIDE WHAT YOU WANT & WRITE IT DOWN: Imagine you have no limitations on what you can Be, Have or Do. Take a sheet of paper, write the word "Goals" at the top, date it, list 10 goals you want to accomplish in the next year. **Be Clear & Detailed!** Unwritten goals are wishes. **Unclear Goal:** "I own a house." **Detailed Goal:** "I own a 3-bedroom, 2 bath, 2400 square foot ranch on a 1 acre mountainside in Hilo Hawaii overlooking the Pacific Ocean with mature palm, banana, and guava trees."

Brian Tracy says to Make Your Goals Personal (use the word I), **Positive** (not negative---not "I lose weight", but "I weigh a healthy fit X-number of pounds by exercising 30 minutes each day."), **and Present Tense** (not "I will", but "I have", "I am", "I weigh", "I earn", "I travel", "I own", etc.).

Step 2: PRIORITIZE YOUR LIST. SELECT YOUR #1 GOAL RIGHT NOW. LIST THE BENEFITS: What will you receive when you achieve the goal. The benefits are your reasons you want to achieve your goal---Answer WHY? The reasons should excite you. **Example: Why do you want to own a home in Hawaii? List the reasons why.** To enjoy spectacular sunrises & sunsets while living in a beautiful place with spectacularly nice weather year-round.

Step 3: LIST THE OBSTACLES TO OVERCOME: Achieving any worthwhile goal will involve challenges and obstacles---list the things you'll need to overcome to achieve your goal so you can prepare yourself in advance. **Example: Finding a job in Hilo, qualifying for a mortgage, finding the house.**

Step 4: LIST THE ADDITIONAL SKILLS AND KNOWLEDGE YOU'LL NEED: Learn more about your goal, what it will take to achieve it and what you'll need to learn. Also find out what skills you'll need to gain in order to achieve your top goal. **Example: Research the market for jobs fitting your qualifications; research the housing prices; apply to pre-qualify for a mortgage** based upon your work and the type of house you're seeking.

Step 5: IDENTIFY THE PEOPLE AND GROUPS TO WORK WITH: What people or organizations can assist you in achieving your goal? **Example: An excellent local Realtor, Mortgage Broker, Job Search Firm.**

Step 6: DEVELOP AN ACTION PLAN: Write out the steps you'll need to take to achieve your goal, an Action Plan: Example: (1) Check out the local job market on LinkedIn or other job web search site, (2) Apply for a mortgage. (3) Find a local Realtor who knows the area and start looking at houses you like.

Step 7: SET A DEADLINE: A goal without a deadline is only a wish. Brian says that a **deadline creates a "subconscious forcing system".** This leads to accountability. And if you miss the date, set another date---but write it down! **ONLY YOU ARE RESPONSIBLE FOR ACCOMPLISHING YOUR GOALS! Example:** I own my beautiful home (described in detail above) by August 18, 2025.

Step 8: TAKE ACTION TOWARD YOUR GOALS EVERY DAY. Launch in the direction of your goals. **Perhaps even make a visual poster with magazine photos of your goals to see daily! Action is Essential!** Remember the Corridor Principle (p.9)---launch, start moving down your chosen path! **To Accelerate Your Goal Achievement: Re-Write your goals each morning and Review each night!**

Chapter 12: Spacecraft Science
"Never settle for mediocrity!" – Gary Ryan Blair from "Everything Counts"

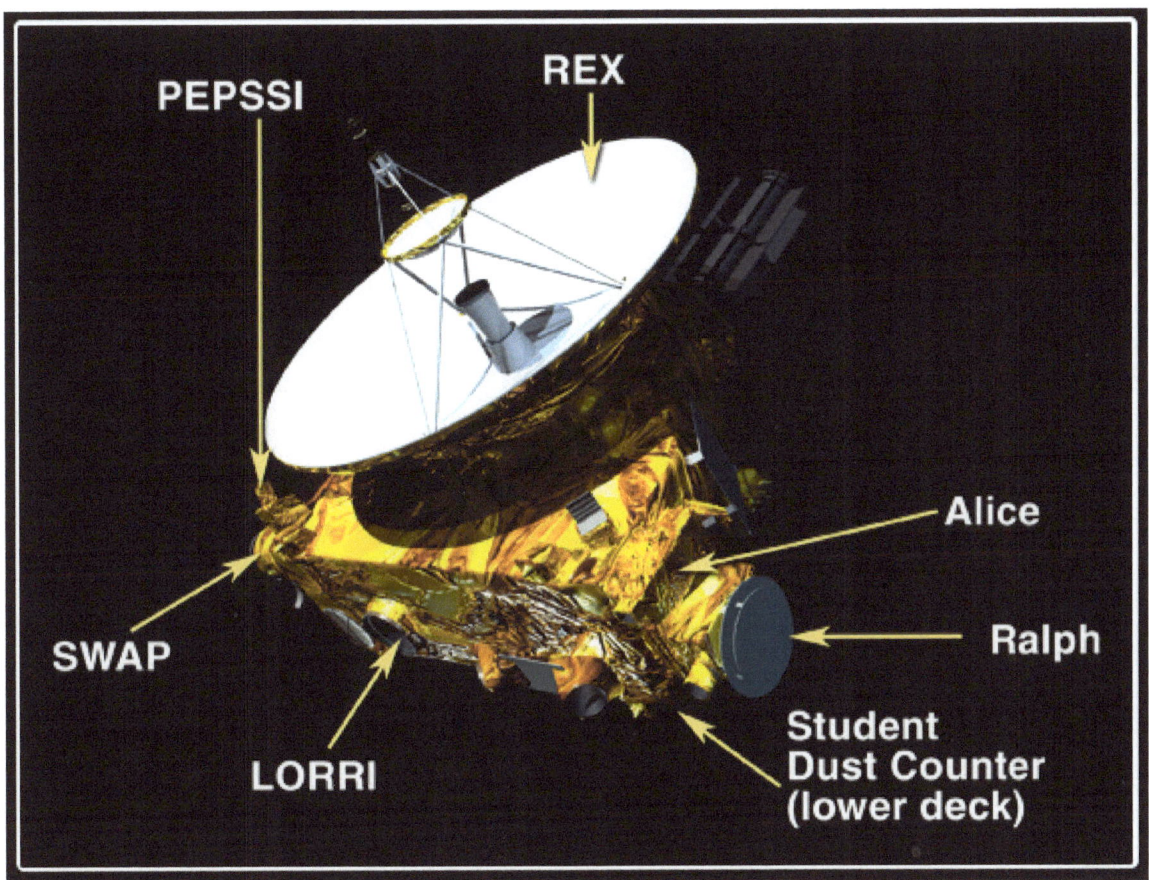

Diagram of New Horizons' Science Payload (Credits: NASA, Johns Hopkins APL, SwRI)

New Horizons carried 7 scientific instruments aboard--- 3 optical, 2 plasma, 1 dust sensor, and a radio science receiver and radiometer---designed to investigate global geology, surface composition, temperature, and atmospheric pressure. It represented power-efficiency---drawing just 28 watts—and miniaturization of electronics. The Radioisotope Thermoelectric Generator (RTG) served as the power source for the extremely cold, low lighting levels at Pluto and beyond in the Kuiper Belt. Solar panels and batteries are ineffective at such distant places. The New Horizons RTG was provided by the **U.S. Department of Energy** and carried approximately 24 pounds (11 kg) of plutonium dioxide---this fuel decays and releases heat at a high rate, which is converted to electrical power.

ALICE - An ultraviolet imaging spectrometer designed to measure the composition and structure of Pluto's atmosphere.

RALPH - The main camera system on New Horizons designed to make maps that show what Pluto and the Kuiper Belt Objects look like. (The Ralph and Alice names come from an old TV show called "The Honeymooners"). **It consists of 8 charge-coupled-devices (CCDs---remember CCDs used by astronomers Jewitt and Luu to discover the 1st KBO beyond Pluto in Chapter 7?),** 3 black and white and 4 color images inside its **Multispectral** Visible Imaging Camera and an infrared mapping spectrometer.

REX - The Radio Science Experiment: an experiment to determine various aspects of the atmospheres of Pluto and Charon during the 2015 flyby---like pressure and temperature, using the X-band radio uplink.

LORRI - A Telescopic Long-Range Reconnaissance Imager: To take highly magnified long-range images.

SWAP - The Solar Wind Around Pluto instrument: used to measure interactions of Pluto with the solar wind from the Sun.

PEPSSI - The Pluto Energetic Particle Spectrometer Science Investigation – used to study the density, composition and nature of energetic particles and plasmas resulting from the escape of Pluto's atmosphere.

SDC - Renamed **Venetia Burney Student Dust Counter** – used to measure concentration of microscopic dust grains produced by collisions among asteroids, comets and Kuiper Belt Objects during New Horizons' long journey. This instrument was designed and built by students at the University of Colorado at Boulder.

For more information, please visit www.pluto.jhuapl.edu/Mission/Spacecraft.php
You can look up the most current Kuiper Belt details & Pluto stats at https://solarsystem.nasa.gov

How does New Horizons send pictures back home?
New Horizons carries high definition **Charge Coupled Device (CCD) Imagers**, large amounts of memory, and a 2.1 meter (83-inch diameter) high gain dish antenna to point toward Earth and send the pictures as 1s and 0s by radio, to be re-assembled by computers on Earth. This includes photos through various color filters. Quite an accomplishment! At the distance of Pluto, and traveling at the **speed of light**, these radio signals take over 4.5 hours to reach Earth!

That's like saying "Hello" by phone to an astronaut at Pluto and waiting 9 hours to hear a round trip reply. That would be a terribly slow conversation, and that's what will happen when people eventually travel to Pluto far in the future. Round trip astronauts travel time to Pluto would be 19 years!

On Earth, 3 roughly equally spaced (120 degrees apart) large dish antenna of NASA's **Deep Space Network** (DSN) provide continuous communications coverage and tracking with spacecraft.
The 3 Global locations of the Deep Space Network around the world: (Credit: NASA, JPL)

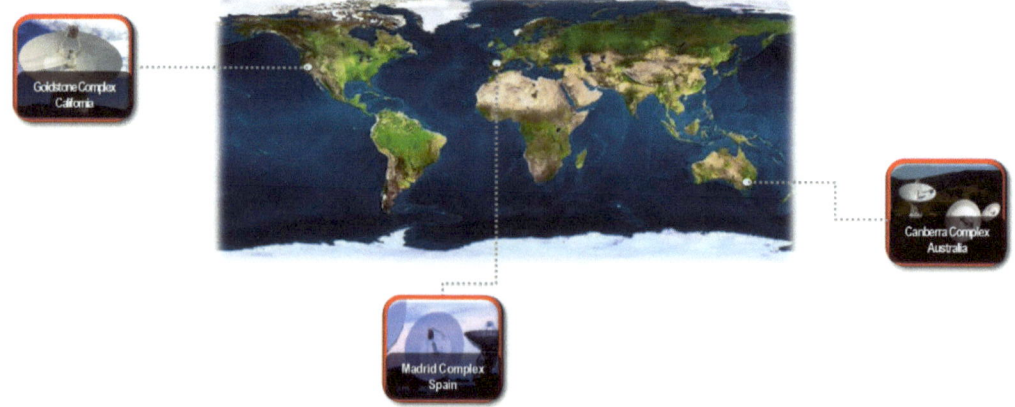

> **Syndicated Columnist Ann Landers** was asked to give what she considered to be the single most useful bit of advice for all humanity: "Expect trouble as an inevitable part of life, and, when it comes, hold your head high, look it squarely in the eye and say, 'I will be bigger than you. You cannot defeat me.'"

Let's Talk about 2 Special Shapes in the World:

70-meter NASA Deep Space Network
Goldstone Antenna in Mohave Desert, CA
(Credit right: NASA)

(1) Parabolic Shape: $y=ax^2$

What makes communications possible using Dish Antennae? (Credit below: The Author)

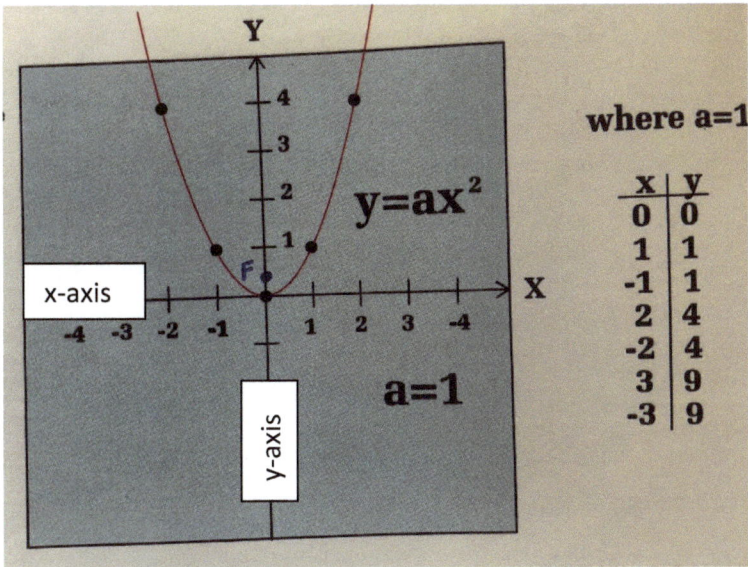

Of course, we need the electronics and software to process the signals.

But did you realize the shape of the parabolic dish is key to its function?

Let's talk math: The equation for a **parabola in 2-dimensions,** with a vertex at the (0,0) point and with the y-axis as the axis of symmetry is $y = ax^2$ where the value of "a" determines dish diameter.

This shape has a <u>**focus F**</u> at y= (1/4)a on the axis of symmetry. **If we spin the parabola about its y-axis of symmetry, we get a 3-dimensional dish antenna shape.** Incoming signals reflect off the inner dish and bounce to the focus. And signals which radiate from the focus, hit the dish surface and are reflected parallel to the axis of symmetry, like parallel laser beams.

This special shape is used in these applications:
- Communications (dish antenna, the Deep Space Network, and New Horizons spacecraft)
- Parabolic microphones (which pick up conversations from across a football field)
- Vehicle headlights and Flashlights (polished reflectors with the bulb at the focus)

What's another special shape which is essential in our world today?

(2) The Shape of an Aircraft Wing!

According to NASA's Glenn Research Center (www1.grc.nasa.gov), <u>Lift</u> is a mechanical force generated by a solid object (i.e. a wing) moving through a fluid (air). **Explore this STEM concept further: Mathematician and scientist <u>Daniel Bernoulli</u> made an interesting discovery about fluid flow, which the Wright Brothers put to practical use for their wing, propeller and control surface designs.** Also applies to helicopter blades.

Wing-tip winglets save fuel $$.
(Credit: Author)

Chapter 13: Success and <u>Bravo!</u>

> "Do not let what you cannot do interfere with what you can do."
> - <u>John Wooden</u>, Basketball Hall of Famer

What did New Horizons Pack for Pluto?
1. Container which holds a small portion of Clyde Tombaugh's ashes and an inscription written by Alan Stern, in honor of the Discoverer.
2. CD-ROM containing the names of > 434,000 people who participated in the "Send your name to Pluto" internet project. What a cool way to inspire hundreds of thousands of people!
3. CD-ROM containing photos and notes from the New Horizons team members.
4. Florida state U.S. Quarter (from which New Horizons was launched).
5. Maryland state U.S. Quarter (where New Horizons was built).
6. A piece of carbon fiber from **SpaceShipOne**, the 1st privately built & piloted spacecraft to reach space (in 2004).
7. Two Small U.S. Flags – one on each side of the New Horizons spacecraft.
8. The 1991 U.S. Postage Stamp proclaiming: Pluto Not Yet Explored. In July 2015, it became Obsolete with the Pluto Flyby!

- **Bravo Percival Lowell and the Lowell Observatory Team!**
- **Bravo Clyde Tombaugh!**
- **Bravo Venetia Burney!**
- **Bravo James Christy!**
- **Bravo NASA, Johns Hopkins University Applied Physics Laboratory & Southwest Research Institute!**
- **Bravo Alan Stern and the New Horizons Team!**
- **Bravo Hubble Space Telescope Team!**
- **Bravo and Thank you New Horizons Team** for sharing the live Pluto Encounter with the world via the internet and for inspiring so many (and future generations) to apply and share their talents with the entire Earth in such Bold Explorations of the Unknown.

In June 2022, at NASA's Outer Planets Assessment Group (OPAG) meeting, Dr. Alan Stern shared that New Horizons was in good condition and its fuel should permit operations through 2040. NASA granted a mission extension to 2025. The team proposed possible heliospheric, astrophysics or planetary science from its unique vantage point within the Kuiper Belt. New Horizons' Adventure Continues! Well Done Team and Tiny Traveler!

In his book "Chasing New Horizons..." Alan Stern summarized the best of the New Horizons Teamwork and Discoveries this way:

"In accomplishing the exploration of Pluto, the New Horizons team set records and achieved many firsts. But more importantly, we think, they demonstrated to the world some of what are the best qualities of humankind: Inquisitiveness, drive, persistence, and the ability to work in teams to achieve something larger than life."

> **From its discovery on Feb 18, 1930 to Pluto's flyby July 14-15, 2015, 85 years had elapsed and transformed Pluto from farthest spec of light to fascinating world, first of the Kuiper Belt Objects!**
> **And just as Clyde Tombaugh felt after looking at the mountains on the Moon through his Uncle Lee's telescope, Alan Stern said that he, too, felt that Pluto was now a place.** (Refer back to top of page 5)

Chapter 14: Lessons Learned, Kuiper Belt Wisdom

"When you have exhausted all possibilities, remember this: You Haven't!"
- **Thomas Edison** (he performed over 10,000 experiments in his goal to invent a successful incandescent light; he did!)

Here are many of the Lessons Learned & <u>Gleaned</u> from the Pluto's Journey, from Pluto's Characteristics, Discovery, Demotion, and the New Horizons' Flyby:

- Be authentic. Be proud of who you are and where you're from and appreciate your uniqueness – Clyde carried his strong Midwestern farming values with him – family, integrity, hard work, perseverance.
- Be bold & willing to walk down your corridor, to change, to become who you want to become
- Reframe Negative Experiences to your benefit. Seek out the Lesson – Pluto was demoted from 9th planet. However, it is now the 1st Kuiper Belt Object of perhaps millions, and the Leader of the 3rd great class of Solar System Objects.
- Remain Humble – remember, you live on a "pale blue dot" when viewed from Pluto.
- Stay Young – It's an Attitude! – People born on Pluto would live their entire lives being less than 1 Pluto Year Old, because Pluto takes 248 Earth Years to orbit the Sun once, its year!
- Face each day with an open heart – Pluto itself has a glacier and icy plane shaped like a heart.
- Prepare, Do your Homework, Research, Then Make a Decision, Commit to a course of action in the direction of your goals, even without credentials, and Take A Leap of Faith – Clyde studied mathematics, the planets, built his own telescopes, and thus qualified himself for the job at Lowell Observatory, then bought a 1-way train ticket to Flagstaff without enough money to return home to Kansas.
- Set yourself apart, Separate yourself from the crowd – Pluto is distinctly different than the other 8 planets.
- Move quietly among the "big boys" but don't become one of them! – Pluto's eccentric orbit takes it closer to the Sun than Neptune for about 20 years of its orbit.
- Maintain your Integrity, and Never Give Up – the New Horizons Team took 26 years from first discussion to actually exploring Pluto.
- Work Hard – Clyde worked winter nights in the cold Lowell Observatory for 13 months
- Be eccentric, Chart your own unique path – Pluto's orbit is tilted about 17 degrees from the plane of travel of the other 8 planets. Don't be afraid to be different!
- Choose a great attitude each day – it's a choice. Remember people reflect back what you radiate out.
- Remain small and agile (quick and responsive) like Pluto – these are competitive advantages in today's world.
- Dare to be Different! And Quietly Lead Your Own Field! - Pluto is the first of perhaps millions of Kuiper Belt Objects in our Solar System---the 3rd great class of Solar System Objects
- Little Things can make a Big Difference – just look at the people inspired by Pluto!
- Never Stand Still – take time to reflect, but maintain momentum toward your goals.
- Be Reflective – Pluto doesn't generate any of its own light, it reflects sunlight
- Use and Share your Talents with the World – as did Clyde's uncle, and as did Clyde with teaching in his later years.
- Keep Cool – The temperatures on Pluto and objects in the Kuiper Belt are extremely cold.

Keeping cool, particularly under pressure is a useful strategy in business and life. It was this repeatedly demonstrated trait during emergencies which led NASA to select Neil Armstrong as the commander for the first manned attempt at landing on the moon during Apollo 11.

- Inspire others – Clyde's Uncle Lee inspired Clyde with astronomy and telescopes. The New Horizons Team inspired 434,000 people to send their names to Pluto, to vote on the names of newly discovered Moons of Pluto, and to share the entire Flyby with the world.
- Persist & Adapt – The Pluto Underground Team proposed 6 missions before NASA finally accepted a mission to robotically explore Pluto.
- Leverage the experience of others to shorten your journey, like Pluto leveraged Jupiter.
- Lead by Example, Lead your field – Percival, Clyde, Venetia, Alan Stern were/are leaders.
- Engage others in your passion – 434,000 people sent their names to Pluto aboard the New Horizons spacecraft, and the New Horizons Team shared their Pluto Flyby with the World via the Internet, including some fanfare with well known personalities like Dr. Brian May of the rock band Queen.
- Remain somewhat mysterious & difficult to detect (by your competition) – like Pluto itself
- It's Up to you! Only you can choose a path and launch toward your goals down the corridor of your chosen path. No one else is going to do it for you or hold your hand. You must take the action and earn the rewards you seek through your own efforts and initiative.
- Plan in advance – Clyde prepared himself & qualified himself for the job at Lowell
- Practice and Rehearse for important events – like New Horizons Team practiced the Pluto encounter years in advance to ensure success at Pluto
- Have Fun & Celebrate Team Successes – The New Horizons Team Celebrated with a "PlutoPalooza" event.
- Have a Primary Partner – Pluto has Charon, plus 4 other Moons.
- Be meticulous about the details – Jim Christy's notice of Pluto's Bump, and Clyde's viewing of photographic points of light.
- Share your discoveries with the world – New Horizons Team did this via a live Flyby internet sharing.
- Use the skills you have – Clyde prepared himself, without formal credentials, then went to Flagstaff, eventually discovering Pluto and thereby creating his own credentials.
- Keep your eyes open – James Christy noticed the bump on the Pluto images which turned out to be Pluto's largest Moon which he named Charon, after his wife.
- Accept responsibility – Clyde accepted the full responsibility of photographing, developing plates, blinking the plates and searching for Pluto, over 7000 hours of hard work.
- Develop a long time perspective – set some long term goals – Pluto's 1 year is 248 Earth years.
- Make daily progress, Move forward daily – Maintain forward momentum.
- Remember, Small is Beautiful! – Like Pluto and its Five Moons!
- Remain Teachable – Clyde kept learning and continuously improving, which kept qualifying him for new opportunities.
- Keep Learning and Continuously Learn! – Remember Einstein's exam? The answers are changing rapidly in most fields. Your formal education cannot carry you for a working lifetime. Remember the AAA, and turn driving time into learning time. You'll put yourself miles beyond most of your competitors.
- Maintain an Attitude of Gratitude and Give Back – Clyde felt grateful for his learning, his opportunities, and his life. He strove to give back to the world with his ongoing sharing of astronomy and his love of learning.
- Keep a Low Mass – stay fit.
- Look for the Best in Others – This one's just a great philosophy!

- Think Yes, not No – The Pluto Underground was told NO by NASA 6 times before acceptance.
- Be Patient - Pluto takes 248 Earth Years to orbit the sun. As of 2023, Pluto has not yet orbited the sun once since the signing of the Declaration of Independence on July 4, 1776!
- Have your Antennae Extended – like the Deep Space Network, always be listening.
- Show others by your actions, not merely your words – the Pluto Not Yet Explored stamp was flown aboard the New Horizons spacecraft to Pluto and Beyond, making it obsolete!
- Be First in the Field – Pluto is first of the new class of Solar System Objects: Kuiper Belt Objects.
- Make a Difference in someone else's life. Inspire them. Share your passion, like Uncle Lee.
- Think Sideways – over, under, around and through---the Ant Philosophy!
- Don't Conform – Pluto is Unique among the 8 other planets.
- Celebrate Fractional Birthdays – 1/248 on Pluto.
- Be open to Change – It's part of daily life. Remember, nothing lasts forever. All records will eventually be broken.
- Get Noticed, eventually. It's okay to "fly under the radar" as long as you like.
- Expand Your Horizons – New Horizons continued its journey to Arrokoth.
- Aim High & Far – like the New Horizons mission.
- Don't Accept Limitations – like Venetia Burney – Age doesn't matter!
- Find another way – follow the Ant Philosophy!
- Never stand still – like Pluto, every other thing in the Universe!
- Create your own Brand – Pluto is unique among the other 8 planets. Capitalize on your strengths.
- Never Settle for Mediocrity – Leave Mediocrity to others! Instead, Dare Mighty Things!
- Know that you're part of something bigger than yourself – the Solar System, The Milky Way Galaxy, The Universe, Humanity, Citizens of Planet Earth.
- Remain curious – this led to a lifetime of work which Clyde loved.
- Do what you love to do. – Clyde loved learning and astronomy and he pursued both for his lifetime.
- Don't accept limitations – Age doesn't matter – just ask Venetia!
- Break the Rules - New Horizons Mission re-wrote the history books about Pluto
- Have Fun, Maintain a Sense of Humor – Laugh – Clyde really enjoyed Puns.
- Act Boldly! Dare Mighty Things! Break the Rules! Be a Leader! – like New Horizons, Like NASA JPL and the Mars Helicopter or JPL's SkyCrane Maneuver!
- Take Initiative – Clyde mailed his sketches to the Lowell Observatory which led to this book!
- Ignore titles, or at least have fun with them – Create your own title! As author Brian Tracy says, "You are the president of your own personal services corporation. And your current employer is your best client!"
- Be Unpopular When Necessary – Re-classifying Pluto as a "Dwarf-Planet" was a good thing.
- Be in the game for the long term - Develop a long time perspective: 9.5 year journey to Pluto.
- Be willing to Act without Credentials---then Create your own credentials, like Clyde.
- Be willing to rethink old beliefs based upon new evidence – science is constantly changing.
- Chart your own unique path – like the New Horizons team which explored Arrokoth.
- Be Resilient and Meet life's challenges Head-On as they come – Clyde and the hail storm
- Ignore Limits – New Horizons was the fastest spacecraft in history! Wait until Star Trek!
- Dare to be different, break the mold! – Think Yoda! New Horizons was the 1st mission to Pluto.
- Act with a Sense of Urgency – Clyde took his work seriously and worked hard daily.
- Be Unstoppable – At Pluto Flyby, New Horizons was travelling about 51,000 MPH, about 30 times faster than a bullet---based upon 1700 MPH for a bullet (51,000/1700=30)!

- Surround yourself with Smart Talented Capable People...and get out of their way! – The New Horizons Team did this.
- Re-write history! Be willing to re-think old beliefs based upon new evidence! – The entire New Horizons mission rewrote the history books about Pluto. Scientific Method does the same.
- Take Responsibility for turning "nothing" into "something" – What do YOU want to create?
- Boldly Go! Make a Difference, give back, engage others in your passion, inspire someone, face challenges head on, be unpopular when necessary, dare to be different, and GO OUT AND DO SOME IMPOSSIBLE THINGS! You are responsible for creating the life you imagine! Go for it!

Lessons from Chapter 1:

Lesson 0.5: Engage others in your passion. Share your passion and your gifts with the world. **Develop an appetite for learning through reading.** Reading daily in your field of interest is the key to becoming highly knowledgeable in that field.

Lesson 1: Learn from the best. Who are the best? Those who have done or who are already doing what you want to do. Seek them out and ask for their help and advice. **Leverage** their experience in order to shorten your path. Then Take Action in the direction of your goals. Your own action or forward motion is the key to getting results.

Lesson 1.5: Share your passion with others, as Uncle Lee did with Clyde about Astronomy and Telescopes. **You don't know who you may be inspiring by sharing your gifts!**

Lesson 2.0: Do what you can with what you have, right where you are. Innovate. Be creative. Maintain a positive optimistic attitude and move forward. Learn all you can and apply your learning and your talents. Clyde learned all he could about telescope building by reading articles, then built several, earning money for his work. He drew sketches of planets using his large home-build telescope, thereby becoming a serious self-educated amateur astronomer. He enjoyed observing planets. He forged a new identity for himself by stepping forward into astronomy.

Lesson 2.5: Expect Obstacles and Adversity on the path toward your goals. Choose and maintain a positive optimistic attitude daily. Keep moving toward your goals.

Reinterpret or reframe your negative circumstances and obstacles---what are they trying to teach you? Look for and glean their lessons.

Remember This: Everything Counts! The good, the bad, the obstacles, and progress. Keep preparing yourself, improving, learning from every experience. Keep moving steadily forward.

Lesson 2.7: Taking action in the direction of your goals is like initiating the automated **launch sequence** of accomplishment.

Lesson 3: Be Authentic, Resourceful and Creative (use the tools and skills you have right where you are) and continue to learn. Clyde was a hard-working farm boy who read about telescopes, wrote to a telescope builder, built his own telescopes from old farm machinery and car parts. Then he drew sketches of Jupiter and Mars from his observations. He continued to read and learn about astronomy. He sent his sketches to the astronomers at Lowell Observatory for their feedback. **His personal efforts led to a job offer, to his becoming the discoverer of a new world, to a university scholarship, to**

meeting his future wife, to his family, to a second career teaching in the military, and to a life unimagined. Be open to such possibilities! Anything is possible!

Lessons from Chapter 2:

Lesson 4: Be Willing to Act Without Formal Credentials! Clyde went to work at a state-of-the-art Observatory based upon nothing more than a passion for astronomy, books he read, his self-taught skills as a telescope builder, and some sketches of the planets he observed. Clyde eventually created his own credentials by having faith to take action in the direction of his goals. **Act, move forward to see doors of Opportunity. Boldly Go! Step forward and take action to become the person you want to become!** Clyde had activated the Corridor Principle, without knowing it by that name.

Lesson 4.5: Get familiar with the Corridor Principle. Dr. Robert Ronstadt of **Babson College's Entrepreneurship** Program in the late 1970s developed and taught a concept called the "**Corridor Principle**". In an article he published, he described the Wright brothers, McDonald brothers, Bill Gates and other entrepreneurs as having something in common. They started earlier ventures and leveraged (there's that word again) and applied their prior experiences to new opportunities that arose toward their new venture. **The Corridor Principle requires that a person take action**---like walking down a corridor or hallway or path (toward a goal which we choose).

Hidden Possibilities and "Doors" of Opportunity only become visible as we start moving along a "corridor" or path toward a goal which we choose, while maintaining a positive, seeking, grateful, expectant attitude. KEY: Doors often led to Unanticipated Successes in areas not originally pursued!

If you stay at the beginning of the hallway or corridor, you cannot see the many doors of opportunity on either side of the hall. **But once you start walking down the corridor, these possibilities and doors of opportunity will become visible to you---they were there at all times, but you'd never have seen them had you not started walking down your chosen path (or corridor). You must be in motion!**

And a positive attitude is essential. In his audio learning program entitled "**The Essence of Success**", author and co-founder of the **Nightingale-Conant Corporation** of Chicago (www.nightingale.com) **Earl Nightingale** said **"Attitude"** is the magic word. He said that the world reflects back to us the attitude that we project to the world. And he further stated that attitude is a choice we make each day. It's a rephrasing of the biblical principal: What you send out, you get back. Or what you sow, you'll reap. In flying terms, a proper wing attitude maintains lift and keeps a plane flying. **So how's your attitude?**

Lesson 5: Don't permit others to define who you are. That's your job! Ignore the noise and chatter around you. It doesn't matter that other people say about you. Continue to focus on your goals. People called Pluto small, a planet, a "dwarf-planet", not a real planet. It didn't matter. Even Clyde once said that whatever Pluto is called, it's still there. Well said Sir!

Lesson 6: Never Give Up! Take a Leap of Faith, Take Action and Go For It when a great opportunity presents itself. Pursue your goals. Clyde left everything behind to pursue his passion for astronomy. Even if you have no formal **credentials** (Clyde had none), take action in the direction of your goals, believe in yourself, persist, find a way to pursue your goals.

Lessons from Chapter 3:

Lesson 6.5: Do whatever it takes---maintaining the fire, clearing the snow---**to get the job done right and in a responsible manner.** People who work in small companies refer to this as "wearing many hats"---hats of various responsibilities, not just one job description.

Lesson 6.9: Prepare yourself in advance for the increased responsibility which will surely come. And when it does, take that responsibility firmly by the reigns and lead the way.

Lesson 7: Work hard. Persevere! Keep moving toward your goals. This is an example of Hard Work and **PERSEVERANCE**---also the name of the newest Mars Rover (the one which carried the 1st helicopter named **Ingenuity** to Mars.

Lesson 7.2: "Do not wait; the time will never be 'just right'. Start where you stand, and work with whatever tools you may have at your command, and better tools will be found as you go along." – **Napoleon Hill**

Lesson 7.5: Be confident in your work. Trust your results but verify.

Lesson 8: Be patient. Success takes time! Hard work and Persistence eventually pay off!

Lesson 9: Be Authentic. Remain humble. Be proud of who you are and where you're from. You are unique in the universe and have an equally unique set of talents and gifts to share with humanity.

Lesson 10: In your work, don't worry about who gets the credit. Support the Team toward the goal. Truth will reveal itself in time.

Lesson 10.5: Create your own credentials through excellence, hard work and by example. Through hard work and persistence and working toward his self-determined goals, Clyde earned his credentials as a world-class astronomer, and earned the formal education he couldn't afford when he began his journey to Flagstaff. **Please note that Lowell himself was a self-taught astronomer too.**

Lessons from Chapter 3.5:

Lesson 11: Age Doesn't Matter! Venetia gave Pluto its name! There are no limits to what you can do! As Nike's motto states: "Just Do it!" **Side Note: See Voyager 1 & 2 Recorded Greetings including 6-year-old Nick Sagan sent to the stars:** Click on right column messages: https://voyager.jpl.nasa.gov/assets/audio/golden-record/english.au

Lesson 11.5: Pay no attention to your age. Instead, pursue your dreams and goals and passions, and success will be yours. Note that the **Voyager** Records launched in 1977 contained a greeting by **6 year old Nick Sagan, Carl Sagan's son..**

Lesson 12: Stay Young! It's an attitude! If people were born on Pluto, they'd spend their entire lives being less than 1 Pluto year old. That's because Pluto's year (time it takes to orbit the sun) is 248 Earth years. So, a 62 Earth year old would be 62/248 = ¼ **year old** in Pluto years! The people of Pluto would probably know a lot more fractions if they celebrated birthdays as often as we do on Earth (every 365 days). **So just for fun...Celebrate Fractional Birthdays! And Develop a Long Time Perspective! Plan some long-term goals decades in the future!**

Lesson from Chapter 3.7:

Lesson 12.5: Become a Life-Long Learner. This is a minimum requirement for success in the 21st century and beyond.

Lesson from Chapter 4:

Lesson 13: Be Thorough. Give your work---even a simple task---the serious care and attention it deserves. Stay alert and observant. You may be on the brink of an amazing discovery! **And Be a Champion (it's an attitude)**---optimistic, curious, creative, forward-focused, ever-learning, and open to new discoveries.

Lesson from Chapter 5:

Lesson 14: Keep your goals in sight, keep making progress and your vision will become clearer.

Lesson from Chapter 6:

Lesson 15: Expect Adversity & Obstacles on the path to your goals. Adapt. Push Forward.
When you set out to accomplish something which has never been done before, obstacles will appear (not **may** appear, but **will** appear!) and people will tell you "NO!" many times. If the goal is worthy and important, and it doesn't harm any other person, then don't let any obstacles stop you! Use them as steppingstones in the direction of your goals!

Lesson 16: The process of scientific discovery is an unending, fluid flow of change and learning.

Lesson from Chapter 7:

Lesson 17: The Law of Cause and Effect governs how the universe works, not just in physics and motion, but in planting, swimming, baking, farming, working, etc.. It is a key to understanding how the world works. Use this Law to achieve your goals.

Recommendations: Check out Dwarf-Planet Ceres which was explored up close by orbiting NASA spacecraft Dawn from March 6, 2015 through October 31, 2018 when it finally ran out of fuel. Learn more about the amazing Dawn spacecraft which visited not only dwarf planet Ceres but large asteroid Vesta within the Asteroid Belt.

> See https://solarsystem.nasa.gov/missions/dawn/in-depth/

Also check out the amazing images and detailed science information about Ceres, including the bright spots indicating salt deposits from a prior ocean which froze (an amazing scientific discovery!).

> https://solarsystem.nasa.gov/planets/dwarf-planets/ceres/in-depth/

There's an excellent 3D rotational movie and Flyover movie of Ceres there too.

Lessons from Chapter 8:

Lesson 18: Prepare well for success. But plan for the worst & for contingencies. Make sure you have emergency plans A, B, C and D.

Lesson 18.4: Leverage the assistance of others to help you take years off your journey toward your goals.

Lesson 18.8: Prepare, Practice, Rehearse for important events or long journeys. Develop back-up or contingency plans in case something goes wrong (like New Horizons' 2nd computer)! Develop emergency plans A, B, and C---but at least A. Think about the Apollo Astronauts who landed on the Moon from 1969-1972---they practiced and rehearsed every part of the mission many times. **How do students prepare for musicals, plays, or sports games?** They practice and rehearse well before the event. That helps determine success!

Lesson 18.9: Celebrate your Successes!

Lesson 19: If you seek, you will find. The more you search, the more you'll find (and learn and discover).
(Photo Credit: www.pinterest.com)

Lessons from Chapter 9:

Lesson 19.1: Ignore Limits!
- **Roger Bannister** ran the sub-4-minute mile when others said it was impossible. His record lasted just 46 days before being broken again---now that people knew it could be done!
- The first two to climb to the top of Mount Everest, highest point above sea level on Earth were **Edmund Hillary** and **Tenzing Norgay** (a beekeeper in New Zealand).
- The Wright Brothers made history on December 17, 1903 with their first powered flight.

  Ignore Limits! They can be and will be broken!

Lesson 20: "Your assumptions are your windows on the world. Scrub them off every once in a while, or the light won't come in." – Isaac Asimov

Lesson 21: In his book, "A Sense of Urgency", author John Kotter recommends these four actions:
1. Accomplish something important every day.
2. Act with a sense of urgency in all you do.
3. Innovate. Look for opportunities, then take daily action on those opportunities.
4. Continuously improve.

Non-fiction Lunar Science Book (Credit: the Author): **"Back To The Moon: Mankind Returns to the Lunar Surface"** published in 2001, 282-pages, full color. (ISBN# 0-9705150-0-6)
See NASA's Artemis Mission!

Lesson 22: We have more in common with others than we may at first realize! Next time, Start looking for what we have in common. Humanity, for example!

Lessons from Chapter 10:

Lesson 23: Improve your vision! Refine your philosophy as you learn more and get closer to achieving your goals.

If necessary, get a better instrument. Get closer! Enlist the help of others---leverage their strengths and experience. Remain flexible, agile, open to change. Have crystal clarity of purpose!

Lesson from Chapter 11:

Lesson 24: Even once you've achieved your primary goals, make sure you develop additional goals. Stay curious and continue to learn and explore!

After the first Lunar Landing of Apollo 11, **NASA quickly realized it needed to help future Moonwalking Astronauts with setting goals beyond their Moon landing missions.** The same will be true for the Return to the Moon and future missions to Mars.

> **New Horizons Mission successfully completed a 13-year Journey---the 1st exploration of the Pluto System and Arrokoth in the Kuiper Belt! A Daring and Mighty Accomplishment! And it's Not Done Yet!**
>
> **I wish you the very best in your adventures on Earth and beyond! – Kevin Caruso**

Relevant Quotes to Think About:

I used 4 sources for the historical descriptions of the People quoted below: https://en.wikipedia.org and www.britannica.com and www.bing.com and www.google.com

"Imagination is more important than knowledge." and "In the middle of every difficulty lies opportunity."
- **Albert Einstein** (German-born theoretical physicist, widely acknowledged to be one of the greatest physicists of all time. He developed the general and special theories of relativity)

"The most effective way to do it, is to do it." and "Never interrupt someone doing what you said couldn't be done."
- **Amelia Earhart** (American Aviation Pioneer and author. First Female Aviator to fly solo across the Atlantic Ocean)

"To see the Earth as it truly is, small and blue and beautiful in that eternal silence where it floats, is to see ourselves as riders on the Earth together."
- **Archibald MacLeish** (American poet and writer, associated with modernist school of poetry)

"The only way to discover the limits of the possible is to go beyond them into the impossible."
- **Arthur C. Clarke** (English Writer, Sir Arthur Charles Clarke, notable for his science fiction and nonfiction).

"Don't you dare underestimate the power of your own instinct."

- **Barbara Corcoran** (American Businesswoman, investor, speaker, syndicated columnist, author, television personality (Shark on TV's "Shark Tank"))

"Well done is better than well said."
- **Benjamin Franklin** (One of the Founding Fathers of the United States of America, leading writer, printer, political philosopher, politician, Freemason, postmaster, scientist, inventor, humorist, statesman, diplomat)

"In a time of rapid change, standing still is the most dangerous course of action." And "You have within you right now, everything you need to deal with whatever the world can throw at you."
- **Brian Tracy** (Canadian American motivational speaker, self-development & success author of over 80 books and audio learning programs)

"There's only one direction you can coast."
- **Brian Tracy**

"I had always taken seriously one of the adages in one of my schoolbooks:
'Do a thing well, or don't do it at all.'"
– **Clyde Tombaugh** (American Astronomer, Discoverer of Pluto)

"When the dream is big enough, the odds don't matter!" and "Don't let Anybody Steal Your Dream!"
– **Dexter Yager** (Built the most successful Multi-Level-Marketing organization in history: Amway.)

"What you must understand is: The determining factors of any kind of success are failures of one kind or another. Refusing to give up is elemental to success."
- **Earl Nightingale** (American Broadcaster, author, speaker, co-founder of Nightingale-Conant Corp)

"My daughter just thinks that all moms fly the Space Shuttle."
- **Eileen Collins** (Retired American Astronaut, Retired United States Air Force Colonel, First Female Pilot and Commander of the Space Shuttle, former military instructor and test pilot).

"No one can make you feel inferior without your consent."
- **Eleanor Roosevelt** (American First Lady, wife of Franklin D. Roosevelt, 32nd President of the Unites States, United Nations Diplomat and Humanitarian).

"Luck is what happens when preparation meets opportunity."
- **Elmer Letterman** (Author)

"When something is important enough, you do it even if the odds are not in your favor." and "I would like to die on Mars. Just not on impact."
- **Elon Musk** (South-African-born American Entrepreneur and businessman, Founder of X-com which later became PayPal, Visionary, Founder, CEO, Chief Engineer at SpaceX, CEO of Tesla, Inc.)

"A smooth sea never made a skillful mariner."
– **English Proverb**

"The sea is dangerous and its storms terrible, but these obstacles have never been sufficient reason to remain ashore…Unlike the mediocre, intrepid spirits seek victory over those things that seem impossible…It is with an iron will that they embark on the most daring of all endeavors…to meet the shadowy future without fear and conquer the unknown."
- **Ferdinand Magellan** (Portuguese Navigator and Explorer who led the first European voyage to circumnavigate the Earth in a single voyage)

"…If they [the ancient philosophers] had seen what we see, they would have judged as we judge."

- **Galileo Galilei** (Italian natural philosopher, astronomer, mathematician who made fundamental contributions to the sciences of motion, astronomy, the strength of materials and to the development of the scientific method. In this quote, he is defending his beliefs before the Inquisition, in the year 1633, that the Earth is not "fixed in the heavens", but instead moves. Galileo was the first person of record to point a telescope skyward---he saw the Moon, phases of Planet Venus, and Discovered Jupiter's great red spot, and Jupiter's 4 largest Moons, named the Galilean Moons in his honor.)

"It is a most beautiful and delightful sight to behold the body of the moon…[It} certainly does not possess a smooth and polished surface, but one rough and uneven, and, just like the face of the earth itself, is everywhere full of vast protuberances, deep chasms, and sinuosities."
- **Galileo Galilei** (in his 24-page book "Sidereus Nuncius" (translated one way as The Starry Messenger), published on March 13, 1610---320 years before Pluto's discovery was announced!)

"People are always blaming circumstances for what they are. I don't believe in circumstances. The people who get on in this world are the people who get up and look for the circumstances they want, and, if they can't find them, make them."
- **George Bernard Shaw** (Irish playwright, critic, political activist, who wrote more than 60 plays, and was awarded the Nobel Prize in Literature in 1925)

"If one advances confidently in the direction of his dreams, and endeavors to live the life which he has imagined, he will meet with a success unexpected in common hours."
- **Henry David Thoreau** (American naturalist, essayist, poet, philosopher)

"Before everything else, getting ready is the secret of success."
- **Henry Ford** (American Industrialist who revolutionized factory production with his assembly line methods and mass production. Founder of the Ford Motor Company, developed the highly successful Model-T car.)

"Failure is only the opportunity to more intelligently begin again."
- **Henry Ford**

"The heights by great men [and women] reached and kept were not attained by sudden flight, but they, while their companions slept, were toiling upward in the night."
- **Henry Wadsworth Longfellow** (American poet and educator, from "the Ladder of Saint Augustine")

"Life begins when you do."
- **Hugh Downs** (American radio and television broadcaster, announcer, programmer, television host, news anchor, TV producer, author, game show host, music composer)

"Engage!" and "Make it so!" and "Things are only impossible until they are not."
- **Captain Jean Luc Picard** (Fictional character & Captain of the 23rd Century Starship U.S.S. Enterprise NCC-1701-D in the television series Star Trek: The Next Generation, played by actor Patrick Stewart, created by Gene Roddenberry and D.C. Fontana, last quote from Episode 16, 'When the Bough Breaks')

"Dare mighty things!"
- **Jet Propulsion Laboratory Motto** (leading center for robotic exploration of the solar system)

"The Definition of Failure: Errors in Judgment repeated every day.
The Definition of Success: A few Simple Disciplines practiced every day." and "Your personal philosophy is the greatest determining factor in how your life works out." and "Reading is essential for those who seek to rise above the ordinary…The habit of reading is a major Stepping stone in the development of a sound philosophical foundation. It is one of the *fundamentals* Required for the attainment of success and happiness."
- **Jim Rohn** (American Author, Speaker, Business Philosopher, last quote from his book "The Five Major Pieces to the Life Puzzle"))

"Go as far as you can see; when you get there, you'll be able to see farther."

- **J.P. Morgan** (American financier, banker, industrialist, head of J.P. Morgan and Co)

"Be a first-rate version of yourself, instead of a second-rate version of somebody else."
- **Judy Garland** (born Frances Ethel Gumm, American Actress, singer, vaudevillian, dancer)

"Don't be casual about your dreams. Kick it up a notch!" and "Make "NO!" your vitamin!"
- **Les Brown** (American motivational speaker, author, former radio DJ, former television host, former member of the Ohio House of Representatives)

"Disciplining yourself to do what you know is right and important, although difficult, is the highroad to pride, self-esteem, and personal satisfaction."
- **Margaret Thatcher** (British Conservative Party Politician, and former British Prime Minister, Europe's first woman prime minister.)

"Always do what is right. It will satisfy some people and astonish the rest."
- **Mark Twain** (Mark Twain is the pen name of Samuel Langhorne Clemens, American humorist, journalist, lecturer, novelist and author of 'Adventures of Huckleberry Finn' (1885)).

"There are no passengers on spaceship earth, we are all crew."
- **Marshall McLuhan** (Canadian philosopher, author)

"Success doesn't come to you... You go to it."
– **Marva Collins** (American Educator who broke with a public school system she found to be failing inner city children and established her own system to foster independence and accomplishment of students.)

"We may encounter many defeats but we must not be defeated."
- **Maya Angelou** (American poet, memoirist, and civil rights activist)

"The starting point of all achievement is desire. Keep this constantly in mind. Weak desires bring weak results, just as a small amount of fire makes a small amount of heat."
- **Napoleon Hill** (Oliver Napoleon Hill (1883-1970). An American self-help author, best known for his 1937 book "Think and Grow Rich", based upon his interviews with Andrew Carnegie and 504 of America's wealthiest businessmen.)

"You only have to solve two problems when going to the Moon:
First, how to get there. Second, how to get back.
The key is: Don't leave until you've solved BOTH problems."
- **Neil Armstrong** (American Astronaut, First Man on the Moon (Apollo 11 Commander), Test Pilot, Educator)

"That's one small step for a man, one giant leap for mankind."
- **Neil Armstrong**

"Just Do It!"
- **Nike** (Trademark phrase of the Nike company, first coined in 1988)

"What I know is, if you do work that you love, and the work fulfills you, the rest will come."
- **Oprah Winfrey** (American talk show host, Interviewer, host and supervising producer of the top-rated award-winning The Oprah Winfrey Show, media leader, philanthropist)

"I think the International Space Station is providing a key bridge from us living on Earth to going somewhere Into deep space."
- **Peggy Whitson** (Retired American Astronaut, First Female Commander of the International Space Station during Expedition 16, biochemistry researcher, former NASA Chief Astronaut, Holds the record for the oldest woman spacewalker, record for the total spacewalks by a woman, oldest female astronaut ever in

space, first female astronaut to command the International Space Station twice, record holder for total accrued time in space of 665 days---equivalent of a round trip to Mars, making her the most experienced astronaut to date)

"Here is the test to find whether or not your mission on Earth is finished:
If you're alive, it isn't!"
- Richard Bach (American writer and author of many 1970's titles, including 'Jonathan Livingston Seagull')

"It is difficult to say what is impossible for us. The dream of yesterday is the hope of today and the reality of tomorrow."
- **Dr. Robert Goddard** (American Engineer, professor, physicist, inventor, American Rocket Pioneer, credited with creating and building the world's first liquid-fueled rocket which he launched on March 16, 1926.)

"There can be no thought of finishing, for 'aiming at the stars', both literally and figuratively, is a problem to occupy generators, so that no matter how much progress one makes, there is always the thrill of just beginning."
- Dr. Robert Goddard

"All adventures, especially into new territory, are scary."
- **Dr. Sally Ride** (American Astronaut, First American Woman in space), member of the team chosen to investigate the Challenger Space Shuttle explosion, Stanford University professor, Founder of Sally Ride Science, promoting Science Technology Engineering Math (STEM) Education for girls and young women.)

"Far better is it to dare mighty things, to win glorious triumphs, even though checkered by failure…
than to rank with those poor spirits who neither enjoy nor suffer much, because they live in a gray twilight that knows not victory nor defeat."
- Theodore Roosevelt Jr. (Teddy Roosevelt, American statesman, conservationist, naturalist, historian, writer, 26th President of the United States from 1901 – 1909)

"There ain't no rules around here. We're trying to accomplish something."
- Thomas Alva Edison (American Inventor and businessman, described as America's greatest inventor, awarded 1093 U.S. Patents and many more global patents for his inventions, including the phonograph, motion picture camera, incandescent electric light)

"We shall not cease from exploration. And the end of all our exploring will be to arrive where we started and know the place for the first time."
- Thomas Stearns Eliot (Poet, essayist, publisher, playwright, literary critic, and editor)

"Stop doing less than Excellent Work!"
- Tom Peters (American writer and speaker on business management, best known for the book "In Search of Excellence" co-authored with Robert H. Waterman Jr.; quote is from his book "The Pursuit of WOW!")

"The difficult can be done immediately. The impossible takes a little longer."
- **U.S. Army Corps of Engineers** (an Engineering formation of the Unites States Army that has 3 primary missions: Engineering, Military Construction, Civil Works.)

"No problem can stand the assault of sustained thinking."
- Voltaire (pen name for Francois-Marie d'Arouet, a French writer, public activist who helped define the 18th century movement called the Enlightenment.)

"When you're curious, you find lots of interesting things to do."
- Walt Disney (American entrepreneur, animator, writer, voice actor, film producer, television host, pioneer of the American animation industry, founder of Disney Brothers Studio, creator of Mickey Mouse in 1928, founder of Disneyland in 1955, The Mickey Mouse Club, Walt Disney World in 1971)

"Always use the word 'Impossible' with the greatest caution." and "It takes sixty-five thousand errors before you are qualified to make a rocket."
- **Wernher von Braun** (German-born American Aerospace Engineering and space architect, rocket scientist, designer of the Saturn V rocket which carried Apollo 11 astronauts to the Moon in July 1969)

"Never give up! Never, Never Give Up!"
– **Winston Churchill** (Sir Winston Churchill, British Statesman, orator, author, Former British Prime Minister from 1940-1945, 1951-1955, he rallied the British people during World War II, leading his country from the brink of defeat to victory.)

"No! Try Not. Do or Do Not. There is no Try."
- **Yoda** (Fictional character & Jedi Master in the Star Wars franchise (now owned by Disney), first appearing in the 1980 George Lucas film 'The Empire Strikes Back'. The quote is Yoda's response to champion character Luke Skywalker when Luke said he would "try" to lift his spacecraft out of the swamp.)

"You can have everything in life you want if you'll just help enough other people get what they want."

- **Zig Ziglar** (American author, businessman, motivational speaker)

What is Success? It's unique to each person. Author Earl Nightingale offered one excellent definition: "Success is the progressive realization of a worthy goal." He also shared: "We become what we think about."

This means you can create the future you want by changing what you think about most of the time, then taking action. If you think & talk about and take action toward your goals most of the time, you'll likely achieve them. AND if you're working (progressing) toward a worthy goal of your choosing, then you're already a success---even before you achieve the goal. The world typically defines success as "after" you achieve the goal. But Earl says you're a success while you're on the journey. YOU ALONE ARE RESPONSIBLE! YOU ALONE DETERMINE YOUR GOALS, WHAT YOU THINK ABOUT, WHAT SUCCESS MEANS TO YOU AND WHAT YOU BECOME! So how do YOU define Success and who do you want to become?

Lowell Observatory's Pluto Discovery Telescope (right showing photographic glass plate holder) **where Clyde Tombaugh re-wrote solar system history, as the New Horizons Team has done again!** (Credit: The Author).

Visit Lowell Observatory in Flagstaff AZ & book an Astronomer-led tour to experience state-of-the-art planetary observations during day or evening! **Be sure to see** Lowell's amazing state-of-the-art Giovale Open Deck Observatory **and the night-time glowing pathways!** See www.lowell.edu

Giant Glossary/Index: Bold Underlined Words Defined

Credits: Sources of Glossary Definitions: For words, www.dictionary.com For people, things, companies or events: www.Google.com & www.Bing.com & www.wikipedia.com & www.britannica.com No copyright is claimed by the author for these definitions or terms from these sources. Invest some time here...there's a lot more learning just reading this Glossary/Index! Enjoy!

Academy of Achievement (pages 23,45): An online museum of living history, highlighting visionaries and pioneers who have helped share our world. See https://achievement.org

Adages (page 12): Noun. Proverbs or short statements expressing a general truth.

A Journey to the Center of the Earth (page 42): Book by science fiction writer Jules Verne.

Alan Boyle (pages 11,15,34): Author of "The Case for Pluto: How a Little Planet Made a Big Difference"; award-winning science writer and space reporter, formerly of nbcnews.com.

Alan Stern (pages 32,34,37,49,65): American Engineer and Planetary Scientist, principal investigator of the New Horizons mission to Pluto and the Kuiper Belt. He has been involved in 24 suborbital, orbital and planetary space missions. Co-Author of the books "Pluto and Charon", and "Chasing New Horizons". Leading New Horizons' continuing mission within the Kuiper Belt.

Albert Einstein (pages 7,74,50): German-born theoretical physicist, widely acknowledged to be one of the greatest physicists of all time. He developed the general and special theories of relativity.

Albion (page 36): First Kuiper Belt Object discovered beyond Pluto, Aug 30, 1992 by David Jewitt and Jane Luu. AKA 1992QB1.

Alden Tombaugh (page 44): Son of Clyde and Patsy Tombaugh.

ALICE (page 62): An ultraviolet imaging spectrometer designed to separate the different wavelength of light and produce an image of the target at each wavelength, used to measure the composition and structure of Pluto's atmosphere.

American Dialect Society (page 3): Founded in 1889, dedicated to the study of the English language, and publisher of the annual Words of the Year list. See www.americandialect.org

Ames Research Center/Bill Keeter (page 50): NASA Ames Research Center in Moffett Field CA, South of San Francisco, conducts world-class research and development in aeronautics exploration technology and science. See https://www.nasa.gov/ames

Andrew Carnegie (page 59): A Scottish-American industrialist and philanthropist. He led the expansion of the American steel industry in the late 19th century and became one of the world's richest Americans. He gave away about $350 Million to charities, foundations and universities, founding libraries, Carnegie Hall, Carnegie Institution for Science, Carnegie Mellon University, Carnegie Museums and more.

Andy Weir (page 35): Author of "The Martian" a novel which became the movie "The Martian" starring Matt Damon.

Annette Tombaugh Sitze (page 34, 45): Daughter of Clyde and Patsy Tombaugh.

Ann Landers (page 63): A penname created by the Chicago Sun-Times newspaper advice columnist Ruth Crowley in 1943 and taken over by Esther Pauline Lederer in 1955.

Ant Philosophy (page 33): Philosophy created by author, speaker and business philosopher Jim Rohn, which states that Ants Never Quit, they keep seeking another way when obstacles block their path. And they prepare for winter all summer.

Antennae (page 64): Plural form on a singular antenna.

Apollo 11 (pages 32, 60,92): First Successful Lunar Landing on July 20, 1969, crewed by Neil Armstrong (1st man on the moon), Edwin "Buzz" Aldrin (2nd man on the moon), and Michael Collins (orbited the moon in the Apollo 11 Command Module).

Apothecary Scale (page 39): A historical balance-based system of mass and volume units used by doctors and pharmacies (apothecaries) for medical recipes and by scientists.

Applied Physics Laboratory (page 34): The Johns Hopkins University Applied Physics Laboratory which solves complex research, engineering and analytical challenges for the nation. They were charged with Operating the New Horizons Spacecraft which explored the Pluto System and then continued exploring into the Kuiper Belt. See https://jhuapl.edu

Aqua-Lung (page 28): A SCUBA diving device co-invented by Jacques Cousteau and Emile Gagnan in 1937. See patent # p.28.

Arizona State College (page 22): Now Arizona State University (ASU). See https://www.asu.edu

Arrokoth (pages 60,65): a small Kuiper-Belt Object which is the most distant object ever explored by a spacecraft, discovered in 2014 by the New Horizons science team using the Hubble Space Telescope. Word from the Powhatan language meaning "sky".

Arthur C. Clarke (page 53): Sir Arthur Charles Clarke (1917-2008), and English writer notable for his science fiction and non-fiction works.

Asaf Hall (page 18): American Astronomer Asaph Hall. Hall discovered the two moons of Mars using the 26-inch telescope at the U.S. Naval Observatory. He named the moons after the mythical Greek gods and brothers: Phobos (fear) and Deimos (terror) upon the suggestion of Henry Madan, an English chemistry teacher, and Uncle of Venetia Burney. The brothers' mythical father was Ares, the god of war, who was known to the Romans as Mars, hence the connection to the red planet.

A Sense of Urgency (pages 53,73): Business Book by author John Kotter.

Asteroids/Asteroid Belt (pages 36, 41): The Asteroid Belt is a doughnut-shaped region in the Solar System between the orbits of Jupiter and Mars that is occupied by many solid irregularly shaped objects of various sizes, generally smaller than planets, called asteroids.

Astronomers (page 3): An astronomer is an expert or student of Astronomy.
Astronomy (page 5): The branch of science which deals with celestial objects, space, and the physical universe.
Astrophysicist (page 48): An astrophysicist is an expert or student of Astrophysics, the branch of astronomy concerned with the physical nature of stars and other celestial bodies and the application of the laws and theories of physics to explain the astronomical observations.
Asymmetry (page 26): The absence of symmetry (the quality of being exactly similar, as a mirror reflection)---in the case of Pluto's Moon Charon, it appeared to James Christy that Pluto had a large bump on one part (an asymmetry), which turned out to be its largest close moon Charon.
Attitude (page 9, 40): A chosen way of thinking or feeling about someone, something, or the world, that is typically reflected in a person's behavior.
Authentic (pages 7, 16, 66): Genuine, of undisputed origin, made or done in the traditional or original way, based on facts, accurate, reliable.
Babson College (page 9): College for entrepreneurship with campuses in Wellesley, Boston and Miami, with undergraduate and graduate schools of business. See https://www.babson.edu
Back to the Moon: Mankind Returns to the Lunar Surface (page 88): A nonfiction science book about the Moon by Author Kevin M. Caruso, electrical engineer, private pilot, Aerospace Education Volunteer for 29 years and former 9-year NASA JPL Solar System Ambassador Volunteer for Illinois.
Barbara Corcoran (page 15): Born March 10, 1949. American Businesswoman, investor, speaker, syndicated columnist, author, television personality (Shark on TV's "Shark Tank")
Benjamin Franklin (page 24): 1706-1790. One of the Founding Fathers of the United States of America, leading writer, printer, political philosopher, politician, Freemason, postmaster, scientist, inventor, humorist, statesman, diplomat.
Big Bang (page 41): A theory which continues to be strongly supported by high-precision astronomical evidence, originally proposed by astronomer Georges Lemaitre in 1927, stating that billions of years ago (currently estimated to be 13.78 billion years ago), the universe started as a single point which exploded, stretched, and expanded to grow as large as it is now.
Bing (page 56): Microsoft Bing (formerly known simply as Bing) is a web search engine owned and operated by Microsoft.
Blink Comparator (page 12): An instrument which permits rapid viewing of two photographs, blinking back and forth between two images taken of the same area of the sky at different times. This permits the operator to more easily spot objects in the night sky that have changed positions. Blink Comparators were obsoleted by modern electronic Charge-Coupled Devices (CCD) sensors that have replaced large photographic plates, and which permit digital storage and analysis by computer software.
Bodleian Library (page 18): One of the oldest libraries in Europe. See www.bodleian.ox.ac.uk
Bravo (page 65): Exclamation used to express approval, particularly when a performer has done something well.
Brera Observatory (page 10): Brera Astronomical Observatory in the Brera district of Milan Italy, built in 1764 by Jesuit astronomer Roger Boscovich.
Brian May (page 48): Born July 19, 1947, an English musician, singer, songwriter, record producer, author, astrophysicist, and lead guitarist of the rock band Queen.
Brian Tracy (pages 7,33,50): Born January 5, 1944. Canadian American motivational speaker, self-development & success author of over 80 books and
audio learning programs. See https://www.briantracy.com
Buckminster Fuller (page 41): R. Buckminster Fuller (1895–1983) was an architect, engineer, geometrician, cartographer, philosopher, futurist, author, futurist, inventor of the famous geodesic dome, and one of the most brilliant thinkers of his time. For more than five decades, he set forth his comprehensive perspective on the world's problems in numerous essays. He wrote, "We are all astronauts on a little spaceship called Earth."
Calculus (page 52): Originally called infinitesimal calculus, the mathematical study of continuous change. It has 2 major branches: differential calculus (concerning instantaneous rates of change, and the slopes of graphical curves) and integral calculus (concerning accumulation of quantities and areas under the graphic curves). Infinitesimal calculus was developed independently in the late 17th century by Isaac Newton and Gottfried Wilhelm Leibniz.
Canali (page 10): The Italian word for "channels".
Canberra (page 63): The capital city of Australia and one of three locations on Earth for the NASA Deep Space Network.
Carl Lampland (page 14): Dr. Carl Otto Lampland (1873-1951). Astronomer who worked at the Lowell Observatory starting in 1902, designing cameras and telescopes for planetary observation. He worked closely with Percival Lowell on photographic exploration of Mars and later, Pluto. After 45 minutes checking his own work Clyde notified Lampland of the Pluto discovery.
Carl Sagan (pages 16,19,44,54,92): 1934-1996. American astronomer, planetary scientist, cosmologist, astrophysicist, astrobiologist, author, and science communicator. He led the effort to send the first physical messages into space aboard Pioneer 10 & 11 spacecraft and Voyagers 1 & 2 (which carried the "Sounds Of Earth" containing 90 minutes of music of cultures around the world and messages from leaders and people of Earth). Dr. Carl Sagan created and hosted the television series "Cosmos", also the name of one of his many books.
Cartography (page 21): The study and practice of making and using maps.
Cassini Spacecraft (page 16): Cassini-Huygens mission was the largest interplanetary spacecraft built by NASA and probe built by the European Space Agency (ESA), which carried 12 instruments, designed to explore planet Saturn, its ring systems and

moons. Cassini was launched October 15, 1997, flew by planet Venus April 25, 1998, Venus again June 24, 1999, Earth August 17, 1999, through the Asteroid Belt December 1999-April 2000, Explored planet Jupiter December 29, 2000, and then Explored planet Saturn by entering orbit July 1, 2004 for 13 years, ending its mission on September 15, 2017.

Cavus (pages 20,56): Irregular steep-sided depressions that don't seem to be impact craters.

Ceres (page 37): Dwarf-planet Ceres is the largest object in the asteroid belt between Mars and Jupiter. It was the 1st asteroid identified and was first thought to be another planet. Discovered by Giuseppe Piazzi in 1801.

Chad Trujillo (page 3,36): Born November 22, 1973, an American astronomer, discoverer of minor planets and co-discoverer with Mike Brown and David Rabinowitz of dwarf-planet Eris, just slightly smaller than Pluto. Computer software expert who specialized in programming automated search software to search for Trans-Neptunian Objects.

Charlene Christy (page 26): Wife of astronomer James Christy for whom Pluto's largest Moon Charon was named.

Charon (pages 27,47,48): Largest Moon of dwarf-planet Pluto, discovered by astronomer James Christy.

Chasing New Horizons (pages 32, 65): A book about the inside look into the epic 1st reconnaissance robotic mission to the Pluto System by Alan Stern and David Grinspoon.

Clyde Tombaugh (pages 4-7, 12, 17,20,43,56,65): 1906-1997. American Astronomer. Discoverer of the then-planet now dwarf-planet Pluto on February 18, 1930.

Clyde Tombaugh: Discoverer of Planet Pluto (pages 22,23,24,65): A book about Clyde Tombaugh by astronomer David H. Levy

Compasses (page 21): A compass is a magnetometer used for navigation and orientation that shows direction relative to the geographic cardinal directions.

Consume (page 10): To absorb all of one's attention and energy, to preoccupy, engross.

Continental Drift (page 41): Continental drift is the hypothesis that the Earth's continents have moved over geologic time relative to each other, and have "drifted" across the oceans, but were once connected like giant puzzle pieces. This concept was first proposed by Abraham Ortelius in 1596. The concept was independently and more fully developed by Alfred Wegener in 1912, but his hypothesis was rejected by many for lack of any proof of a means of motion. Arthur Holmes later proposed mantle convection as the means for motion. The idea of continental drift has since been absorbed into the theory of plate tectonics, which explains that the continents move by riding on plates of the Earth's lithosphere. This theory is now accepted as the evidence supports it---that's the Scientific Method at work!

Contingency (page 44): Noun. A future event which has a probability of happening, but which cannot be predicted with certainty. A provision for an unforeseen event or circumstance---a "plan B", being prepared for the worst, but expecting the best.

Corridor Principle (pages 8,9,59,61,70): A concept developed and taught by Dr. Robert Ronstadt of Babson College in the late 1970s which states that one can leverage their prior experiences and choose to pursue a path toward their goal, and by moving toward their goal, it's like a corridor in which doors of opportunity will become visible only as you move down the corridor.

Cosmos (page 54): Book by famous astronomer Carl Sagan (1934-1996).

Credentials (page 11): Noun. A qualification, Achievement, Certificate, Diploma, Training, Knowledge, Experience, Formal Education to indicate that a person is qualified for something.

Criteria (page 37): Noun. Plural form of Criterion. A principle or standard by which something may be judged or decided. A qualification or measure, gage, scale, indicator, guideline.

Daniel Bernoulli (page 64): Daniel Bernoulli (1700-1782) was a Swiss mathematician and physicist and one of the many prominent mathematicians in the Bernoulli family. He is particularly remembered for his applications of mathematics to mechanics, especially fluid mechanics, and for his pioneering work in probability and statistics. The Bernoulli's principle is named in his honor: an example of the conservation of energy, which describes the mathematics of the mechanism underlying the operation of two important technologies of the 20th century: the carburetor and the airplane wing.

David Jewitt (page 3,36,62): Born 1958. British-American astronomer who studies the Solar System, especially its minor objects. Professor of Astronomy at University of California. Best known with Jane Luu for being the first people to discover 1992QB1, Albion, an object beyond Pluto and Charon in the Kuiper Belt, thus confirming the existence of the Kuiper Belt.

David H. Levy (pages 22,23,24,37): Born May 22, 1948. A Canadian amateur astronomer, science writer, and discoverer of comets and minor planets who co-discovered Comet Shoemaker-Levy 9 in 1993, which broke apart and collided with planet Jupiter in 1994 as witnessed by the Hubble Space Telescope.

David L. Rabinowitz (page 3,36): Born 1960. An American astronomer and discoverer of minor planets and researcher at Yale University. He, together with Mike Brown and Chad Trujillo discovered Eris, a "dwarf-planet" slightly smaller than Pluto. He has built CCD cameras and written software to improve the detection process for near-Earth objects and Kuiper Belt Objects. He has assisted in the detection of distant solar system objects, supernovae and quasars, helping better understand the origin and evolution of the solar system and dark energy driving the expansion of the universe.

Dawn (pages 36,72): NASA spacecraft, orbited asteroid Vesta, Dwarf-Planet Ceres https://solarsystem.nasa.gov/missions/dawn

Deep Space Network (pages 63,64): NASA's global communications network comprising 3 primary worldwide sites, each approximately 120 degrees apart around the Earth (in California, Spain, Australia) which provides continuous communications coverage with spacecraft as the Earth rotates on its axis.

Deimos (page 18): In Greek mythology, the god of terror & dread, and brother of **Phobos** (panic & fear) and the name given to the smaller of Mars' two moons. Both moons were discovered and named by **American Astronomer Asaph Hall** in August 1877 using the 26-inch telescope at the **U.S. Naval Observatory**. The names Phobos and Deimos were chosen by Hall based upon a suggestion by **Henry Madan**, an English chemistry teacher, and Uncle of **Venetia Burney**. Their mythical father was Ares, the

god of war, who was known to the Romans as Mars, hence the connection to the red planet.
Demotion (page 3): Noun. A reduction in rank or status, ousting, removal, unseating, dethronement, dismissal, removal.
Dexter Yager (page 34): Dexter Yager (1939-2019) built the most successful Multi-Level-Marketing organization in history: Amway. Author of the brief book "Don't Let Anyone Steal Your Dream"
Digress (pages 16,24): Verb. To leave the main subject temporarily in speech or writing.
Dinsmore Alter (page 22): Dr. Dinsmore Alter (1888-1968). An American astronomer, meteorologist, and United States Army officer, known for his work with the Griffith Observatory and his creation of a Lunar Atlas.
Double-Planet (page 26): In astronomy, an informal tern used to describe two planets that orbit each other about a common center of mass that is not located within the interior of either planet. The formal term is "binary system". The term is not officially recognized by the International Astronomical Union (IAU) and is therefore not an official classification.
Dwarf-Planet (pages 3,42): In astronomy, a "dwarf-planet" is a planetary-mass object that does not dominate its region of space (as a planet does) and is not a satellite. It directly orbits the Sun and is massive enough to crush itself into a spheroid shape (hydrostatic equilibrium) but has not cleared the neighborhood of its orbit of similar objects. Pluto is an "dwarf-planet" as are asteroid Ceres and Kuiper Belt Objects Makemake, Haumea, and Eris.
Earl Nightingale (page 9): American Broadcaster, author, speaker, co- founder of Nightingale-Conant Corporation
Earth (page 3): Planet Earth, our home world, and 3rd planet from the Sun in our Solar System.
Economic (page 5): Adjective. Relating to economics or the economy, justified in terms of profitability.
Edgeworth-Kuiper Belt (page 56): Also called the Kuiper Belt, named after Dutch astronomer Gerard Kuiper---he theorized that is where most of the periodic comets originate, though he did not predict its existence. A flat ring of icy small objects like periodic comets orbiting the Sun beyond the Orbit of Neptune---from about 30 astronomical units (AUs) to approximately 47-50 AUs. Thought to be like the asteroid belt, but far larger: perhaps 20 times wider and 20-200 times as massive. It contains remnants from the formation of the Solar System. Irish astronomer Kenneth E. Edgeworth speculated in 1943 that the Solar Systems' small bodies extended beyond Pluto. Kuiper developed a stronger case in 1951. In 1950, Dutch astronomer Jan Oort proposed the existence of a much more distant spherical reservoir of icy bodies, now called the Oort cloud, from which comets are continuously replenished.
Edmund Hillary (pages 51,73): Sir Edmund Percival Hillary (1919-2008). A New Zealand mountaineer, explorer, and philanthropist. On May 29, 1953, Hillary and Sherpa mountaineer Tenzing Norgay became the first climbers confirmed to have reached the summit of Mount Everest. They were part of the ninth British expedition to Everest, led by John Hunt.
Edwin Hubble (page 23): Edwin Powell Hubble (1889-1953). An American astronomer. He played a crucial role in establishing the fields of extragalactic astronomy and observational cosmology. Hubble proved that many objects previously thought to be clouds of dust and gas and classified as "nebulae" were galaxies beyond the Milky Way. He used the strong direct relationship between a classical Cepheid variable's luminosity and pulsation period for scaling galactic and extragalactic distances. Hubble provided evidence that the recessional velocity of a galaxy increases with its distance from the Earth, a property now known as "Hubble's law", even though it had been both proposed and demonstrated observationally two years earlier by Georges Lemaître. The Hubble–Lemaître law implies that the universe is expanding. A decade before, the American astronomer Vesto Slipher had provided the first evidence that the light from many of these nebulae was strongly red-shifted, indicative of high recession velocities. Hubble's name is most widely recognized for the Hubble Space Telescope, which was named in his honor, with a model prominently displayed in his hometown of Marshfield, Missouri.
Eileen Collins (page 75): Born 1956. Retired NASA astronaut and U.S. Air Force Colonel. Former military instructor and test pilot. First female pilot and first female Commander of the Space Shuttle.
Electromagnetic Spectrum (page 41): The electromagnetic spectrum is the range of frequencies (the spectrum) of electromagnetic radiation and their respective wavelengths and photon energies. The electromagnetic spectrum covers electromagnetic waves with frequencies ranging from below one hertz to above 10^{25} hertz, corresponding to wavelengths from thousands of kilometers down to a fraction of the size of an atomic nucleus. This frequency range is divided into separate bands, and the electromagnetic waves within each frequency band are called by different names; beginning at the low frequency (long wavelength) end of the spectrum these are: radio waves, microwaves, infrared, visible light, ultraviolet, X-rays, and gamma rays at the high-frequency (short wavelength) end. The electromagnetic waves in each of these bands have different characteristics, such as how they are produced, how they interact with matter, and their practical applications. The limit for long wavelengths is the size of the universe itself, while it is thought that the short wavelength limit is in the vicinity of the Planck length. Gamma rays, X-rays, and high ultraviolet are classified as ionizing radiation as their photons have enough energy to ionize atoms, causing chemical reactions.
Elmer Letterman (page 8): An American salesman and author of the book "The Sale Begins When The Customer Says No"
Elon Musk (pages 11,75): Born 1971. South African-born American Entrepreneur who cofounded PayPal and formed SpaceX and was one of the first significant investors and CEO of Tesla. Successfully launched crewed missions to the International Space Station. Wants to fly people to Mars. Selected by NASA to land the next crew on the Moon. Lands rocket boosters back onto their launch pads or on moving barges in the ocean! Very Impressive! Way to go SpaceX!
Entrepreneurship (page 9): Noun. The activity of setting up a business or businesses, taking on financial risks to make a profit.
Eris (pages 3,31,36,37): Eris is the Greek goddess of strife and discord. Her Roman equivalent is Discordia. Eris the "dwarf-planet" was discovered on October 21, 2003 by Michael Brown, Chad Trujillo, and David Rabinowitz using the Palomar Observatory. Eris is a little smaller than Pluto but 3 times farther from the Sun. Because it was first thought to be larger than

Pluto, its discovery highlighted the Great Planet Debate in the scientific community which led to the International Astronomical Union's decision in 2006 to define the word "planet". Pluto, Eris and other similar objects are now classified as "dwarf-planets". See https://solarsystem.nasa.gov/dwarf-planets/eris/in-depth/

ESA (page 29): Acronym for the European Space Agency

Essence of Success (pages 9,70): Book and audio learning program about Success by Earl Nightingale.

Extra-Solar Planets (page 41): Also called exoplanets. These are planets outside of our Solar System. The first confirmation of detection occurred in 1992. As of April 2021, there are 4704 confirmed exoplanets in 3478 other solar systems.

Everything Counts! (page 62): A book by Gary Ryan Blair.

Evolution (page 41): Noun. The process by which different kinds of living organisms are thought to have developed and diversified from earlier forms during the history of the Earth. The gradual development of something, especially from a simple to a more complex form.

EPCOT (page 58): Acronym for Walt Disney World Florida's Experimental Prototype Community Of Tomorrow.

Fail-safe (pages 15,44): Adjective. Incorporating some features for automatically counteracting the effect of an anticipated possible source of failure. The images of Pluto and Pluto and Charon transmitted back to Earth as New Horizons spacecraft approached to Pluto system, prior to its close flyby, ensured that scientists would have some worthwhile data and photos of Pluto and Charon should the New Horizons spacecraft be destroyed during the close flyby by some space debris. These images were fail-safe images.

Falconer Madan (page 18): 1851-1935. Librarian of the Bodleian Library of Oxford University and grandfather of Venetia Burney who named Pluto.

Federal Aviation Administration (pages 5,6): Also known as the FAA. The largest modern transportation agency and a government body of the United States with powers to regulate all aspects of civil aviation in the nation as well as over its surrounding international waters.

Flyby (page 3): Noun. A flight past a point, especially the close approach of a spacecraft to a planet or moon for observation, without going into orbit about that planet or moon. The New Horizons mission to Pluto was a Flyby mission.

Focus (page 64): Noun. The center of interest or activity.

Franklin Institute (page 43): Also, The Franklin Institute Science Museum is in Philadelphia, PA. See https://www.fi.edu

Galaxies (page 41): A galaxy is a gravitationally bound system of stars, stellar remnants, interstellar gas, dust, and dark matter. The word galaxy is derived from the Greek word "galaxias" which literally means "milky", a reference to the Milky Way. Galaxies range in size from dwarfs with just a few hundred million (10) stars to giants with one hundred trillion (10) stars, each orbiting its galaxy's center of mass.

Galileo Galilei (page 76): Italian natural philosopher, astronomer, mathematician who made fundamental contributions to the sciences of motion, astronomy, the strength of materials and to the development of the scientific method. In this quote, he is defending his beliefs before the Inquisition, in the year 1633, that the Earth is not "fixed in the heavens", but instead moves. Galileo was the first person of record to point a telescope skyward---he saw the Moon, phases of Planet Venus, and Discovered Jupiter's great red spot, and Jupiter's 4 largest Moons, named the Galilean Moons in his honor.

Gary Ryan Blair (page 62): Author of the book, "Everything Counts!"

General George S. Patton, Jr. (page 38): 1885-1945. A general in the United States Army who commanded the Seventh United States Army in the Mediterranean theater of World War II and the Third United States Army in France and Germany after the Allied invasion of Normandy in June 1944.

Geometry (page 21): Noun. The branch of mathematics concerned with the properties and relations of points, lines, surfaces, solids, and higher dimensional analogs.

George Bernard Shaw (page 76): Irish playwright, critic, political activist, who wrote more than 60 plays, and was awarded the Nobel Prize in Literature in 1925)

Giovanni Virginio Schiaparelli (page 10): 1835-1910. An Italian astronomer and senator whose reports of groups of straight lines on Mars touched off much controversy on the possible existence of life on Mars.

Giuseppe Piazzi (page 36): An Italian Catholic priest of the Theatine order, mathematician, astronomer. He established an observatory in Palermo. He discovered the first asteroid, now "dwarf-planet", Ceres on January 1, 1801. Ceres was at first thought to be a new planet.

Gleaned (page 66): Past tense of Glean. Verb. To extract information from various sources, to collect gradually, bit by bit.

Goal Setting (page 61): Goal. Noun. The object of a person's ambition or effort, an aim or desired result. Goal Setting is establishing and defining goals, writing them down, setting an end date, and making plans to accomplish the goals by the specified date.

Goldstone (page 63): Goldstone is one of three sites around the world known as the Deep Space Network, established to provide the ability to communicate with spacecraft continuously as the Earth rotates.

Google (page 56): Google LLC is an American multinational technology company that specializes in Internet-related services and products, including their Google search engine.

Gottfried Wilhelm Leibniz (page 52): 1646-1716. A German philosopher, mathematician, scientist, diplomat. Co-credited with the independent creation of a new field of mathematics: infinitesimal calculus. Isaac Newton is also credited for this.

Gravity (pages 41,64): Also Gravitation. The natural phenomenon by which all things with mass or energy, including planets, stars, galaxies and even light are brought toward one another. On Earth, gravity gives weight to physical objects. The Moon's gravity causes the ocean tides. There's a tale of an apple falling (due to gravity) onto the head of Isaac Newton sitting under an apple tree, which led to his developing the equation for the force of gravity acting on two masses over a distance.

Haumea (page 37,38): Named after the Hawaiian goddess of fertility, this Kuiper Belt Object is a rapidly spinning football-shaped "dwarf-planet" located beyond Neptune's orbit. Two teams claim credit for discovery---Sierra Nevada Observatory in Spain and at team of Mike Brown of Caltech at the Palomar Observatory, David Rabinowitz of Yale University, and Chad Trujillo of Gemini Observatory in Hawaii.

Helen Keller (page 29): Helen Adams Keller (1880-1968). An American author, disability rights advocate, political activist and lecturer. Born in Alabama, she lost her sight and hearing after an illness at the age of 19 months. She then communicated primarily using home signs until the age of 7 when she met her first teacher and life-long companion Anne Sullivan who taught her language, including reading and writing. She also learned how to speak and to understand other people's speech using the Tadoma method. Attended Radcliffe College of Harvard University becoming the first deaf and blind person to earn a Bachelor of Arts degree. She worked for the American Foundation for the Blind and traveled to 39 countries advocating for those with vision loss. Helen Keller wrote 14 books.

Henry Ford (pages 4,76,61): 1863-1947. American industrialist, businessman, founder of the Ford Motor Company who revolutionized factory production with his assembly-line methods. He converted the automobile from an expensive curiosity into an accessible affordable means of transportation.

Huffman Prairie Field (page 6): An 84-acre cow pasture outside of Dayton, OH, now a National Landmark adjacent to Wright Patterson Air Force Base, where the Wright Brothers learned to fly with precision and for longer distances. There they invented a take-off track raised above the prairie grass with a catapult using a 1600-lb weight to launch their plane into the air, independent of the wind. Wilbur Wright called this a "starting device". At the time, airports didn't exist, flight schools didn't exist, simulators didn't exist, pilots didn't exist.

Hydra (pages 30, 48): The outermost moon of Pluto, orbiting beyond Kerberos. Discovered June 15, 2005. Hydra is approximately 32 miles (51 km) across its longest dimension, 2nd largest moon of Pluto. Named after the Hydra, a 9-headed underworld serpent in Greek mythology.

Hydrostatic Equilibrium (page 38): In fluid mechanics, hydrostatic equilibrium or hydrostatic balance is the condition of a fluid or plastic solid at rest. This occurs when external forces such as gravity are balanced by a pressure-gradient force. It is the description of the process by which the gravity of sufficiently massive objects crushes them into a spherical shape.

Hubble Space Telescope (pages 29,65): An Earth-orbiting astronomical observatory launched aboard and deployed by the Space Shuttle in 1990. The telescope's high-resolution images are far better than can be obtained from the Earth's surface.

Hugh Downs (page 15,76): 1921-2020. An American radio and television broadcaster, announcer, programmer, television host, news anchor, TV producer, author, game show host and music producer.

Infrared Light (pages 30,41): Electromagnetic radiation with wavelengths longer than those of visible light, and therefore invisible to the human eye. Discovered in 1800 by Sir Frederick William Herschel, who also discovered Uranus in 1781.

Ingenuity (pages 13,24): Name of the Mars Helicopter which first flew on Mars 4-20-21, affectionately known as "Ginny"

International Astronomical Union (pages 3,20,37,38): Founded in 1919, the IAU's mission is to promote and safeguard the science of astronomy in all its aspects, including research, communication, education and development through international cooperation. The IAU is responsible for officially naming celestial bodies and features on other worlds.

Isaac Asimov (page 53): 1920-1992. American writer and profession of biochemistry at Boston University. Known for his works of science fiction and popular science. Wrote or edited more than 500 books, and an estimated 90,000 letters and postcards.

Isaac Newton (page 38,39,52): Also see Sir Isaac Newton.

Jack Welch (page 38): Former CEO of the General Electric Company, business lecturer, author.

Jacques Yves Cousteau (page 28): (1910-1997) Oceanographer, Researcher, Filmmaker, undersea explorer, author, inventor.

James Christy (pages 26, 27, 48): Astronomer who discovered Pluto's largest Moon Charon while working at the U.S. Naval Observatory near Flagstaff, AZ. He named the moon after his wife Char (Charlene).

Jane Luu (page 3,36,37,62): Dr. Jane X. Luu. Born July 1963. Vietnamese American astronomer and defense systems engineer. Awarded the Kavli Prize. Co-discoverer on August 30, 1992 with David Jewitt of the first Kuiper Belt Object beyond Pluto.

JHUAPL (pages 32,44,49-57,65): Acronym for Johns Hopkins University Applied Physics Laboratory.

Jet Propulsion Laboratory (JPL) (pages 32,44,49-57,65): NASA Jet Propulsion Laboratory. See https://www.jpl.nasa.gov The Jet Propulsion Laboratory (JPL) is a federally funded research and development center and NASA field center in the city of La Canada Flintridge (with a Pasadena mailing address) in California. Founded in the 1930s, JPL is owned by NASA and managed by the nearby California Institute of Technology (Caltech). Its primary function is the construction and operation of planetary robotic spacecraft, though it also conducts Earth-orbit and astronomy missions.

Jim Rohn (pages 33,38): Emanuel James Rohn (1930-2009), professionally known as Jim Rohn. An American entrepreneur, author, business philosopher and motivational speaker. See https://www.jimrohn.com

John F. Kennedy (page 32): John Fitzgerald Kennedy (1917-1963). An American politician who served as the 35th President of the United States of America from 1961 until his assassination in 1963. JFK initiated the Apollo Lunar Landing Program. See https://en.wikipedia.org/wiki/John_F._Kennedy

Johns Hopkins University Applied Physics Laboratory (JHUAPL) (pages 43,, 47, 48, 49, 50): A not-for-profit university-affiliated research center in Maryland. Employs 7200 people in 2020. The lab serves as a technical resource for the Department of Defense, NASA and other government agencies.

John Kotter (page 53): Author, award-winning business and management thought leader, business entrepreneur and Harvard Professor. Author of the book "A Sense of Urgency". See https://www.kotterinc.com/team/john=kotter

John Wooden (page 65): 1910-2010. An American basketball payer and coach. He won ten National Collegiate Athletic Association (NCAA) national championships in a 12-year period as head coach for the UCLA Bruins.

J.P. Morgan (page 60): John Pierpont Morgan (1837-1913). An American financier and banker who dominated corporate finance on Wall Street throughout the Gilded Age. Head of the banking firm that became J.P. Morgan and Co.

Jules Verne (page 42): 1828- 1905. A French novelist, poet, and playwright. In collaboration with publisher Pierre-Jules Hetzel, led to the creation of best-selling adventure novels: Journey to the Center of the Earth (1864), Twenty Thousand Leagues Under the Seas (1870) and Around the World in Eighty Days (1872).

Jupiter (pages 3,45,46): Planet Jupiter, the largest planet---a gas giant---and 5th planet from the Sun in our Solar System.

Kepler's Laws (page 22): Three theorems describing orbital motion developed and published by Johannes Kepler between 1609 and 1619. The first law states that the planets move in elliptical orbits with the sun at one focus. The second law states that the radius vector of a planet sweeps out equal areas in equal times. The 3rd law relates the distances of the planets from the sun to their orbital periods. Kepler's laws modified the heliocentric theory of Nicolaus Copernicus.

Kerberos (pages 30,48): One of 5 moons of Pluto, discovered on June 28, 2011 by Astronomer and New Horizons team member Mark Showalter using the Hubble Space Telescope. The discovery was confirmed by additional imaging by Hubble then announced on July 20, 2011. The name Kerberos (also spelled Cerberus) is a mythical 3-headed dog guarding Pluto's realm.

Kevin Caruso (pages 7,33,39,74): Electrical Engineer, private pilot, proud father, Quality Manager, Author of this book and the book "Back to the Moon: Mankind Returns to the Lunar Surface"

Kuiper Belt (page 25,36): A circumstellar disc in the outer Solar System, extending from the orbit of Neptune at 30 astronomical units (AU) to approximately 47-50 AU from the Sun. It is like the asteroid belt but far larger – perhaps 20 times as wide and 20–200 times as massive. Like the asteroid belt, it consists mainly of small bodies or remnants from when the Solar System formed.

Kuiper Belt Object (KBO) (pages 3,36,37,38,41): Smaller celestial objects, many bits of rick and ice, comets, and "dwarf-planets" which orbit the Sun in a large flattened zone beyond the orbit of Neptune. Also called Trans-Neptunian Objects (TNOs). Pluto is the largest and 1st of the KBOs in our Solar System---often referred to as the 3rd great class of Solar System Objects---after the 1st class of rocky inner planets (Mercury, Venus, Earth, and Mars), and 2nd class of Gaseous Giants (Jupiter, Saturn, Uranus, and Neptune).

Launch Sequence (page 7): The sequence of operations or steps and verifications (often performed in an automated manner by computer control) which must occur for a rocket to safely launch into space.

Law of Cause and Effect (pages 33,39,40): The Law of Cause and Effect states that every cause has an effect and every effect has a cause. The suggests that the universe is always in motion. The great news is that this law describes an orderly universe in which you have control over many of the causes (your thoughts are causes) in your life, whereas the effects usually occur without any assistance from you, in response to your causes. This is a variation of Sir Isaac Newton's 3rd Law of Motion: For Every Action, There is an Equal and Opposite Reaction.

Leverage (pages 5, 55): Noun. The exertion of force by means of a lever or an object used in the manner of a lever. Also, the use of maximum advantage, as in leveraging the knowledge and experience of others by listening to experienced people, reading their published books and listing to their audio learning programs or seminars.

Les Brown (page 34): Author and Motivational Speaker.

Lift (pages 9,64): One of the 4 Forces of Flight: Lift, Gravity, Thrust, Drag. Lift is the force that acts at a right angle (up) to the direction of motion through the air. Lift is created by the differences in air pressure above and below a wing or airfoil, caused the air flow over the "special shape" of the Wing. See Daniel Bernoulli.

LORRI (pages 62,63): Long Range Reconnaissance Imager aboard the New Horizons spacecraft, a long-range telescopic imaging camera used for taking approach images from a great distance.

Lowell Observatory (pages 4,7,8,10,12,14,15,17,65,66): Lowell Observatory is an astronomical observatory in Flagstaff, Arizona, United States. Lowell Observatory was established in 1894 by Percival Lowell, placing it among the oldest observatories in the United States, and was designated a National Historic Landmark in 1965. In 2011, the Observatory was named one of "The World's 100 Most Important Places" by TIME magazine. At Lowell Observatory, American Astronomer Clyde Tombaugh discovered then-planet Pluto on February 18, 1930. In 2006, Pluto was "Pluto-ed", that is, reclassified as a "dwarf-planet".

Madrid (page 63): The capital and most populated city of Spain. The 2nd largest city in the European Union. Madrid is also one of the 3 locations of the NASA Deep Space Network for spacecraft communications.

Magnitude (page 14): Noun. The great size or extent of something. In astronomy, the degree of brightness of a star. For example, a star with an apparent magnitude of six is barely visible to the naked eye.

Makemake (page 38): A "dwarf-planet" or Kuiper Belt Object (KBO) discovered on March 31, 2005 by Michael Brown, Chad Trujillo and David Rabinowitz at the Palomar Observatory.

Malin Space Science Systems (MSSS) (page 24): Established in 1990, Malin Space Science Systems (MSSS) designs, builds, and operates space camera systems for government/commercial aerospace missions including Mars. See www.msss.com

Marc Buie (page 60): Marc William Buie is an American astronomer and prolific discoverer of minor planets, who works at the Southwest Research Institute in the Space Science Department. He formerly worked at Lowell Observatory in Flagstaff, Arizona, and the Sentinel Space Telescope Mission Scientist for the B612 Foundation, which is dedicated to protecting Earth from asteroid impact events. Marc was a team member on the New Horizons mission to Pluto and beyond. Using the Hubble Space Telescope, Marc discovered KBO Arrokoth, which New Horizons flew past on January 1, 2019.

Margaret Thatcher (page 77): 1925-2013. A British stateswoman who served as Prime Minister of the United Kingdom from 1979 to 1990 and Leader of the Conservative Party from 1975 to 1990. She was the longest-serving British prime minister of the 20th century and the first woman to hold that office.

Marie Curie (page 52): 1867- 1934. A Polish and naturalized-French physicist and chemist who conducted pioneering research on radioactivity. She was the first woman to win a Nobel Prize, the first and the only woman to win the Nobel Prize twice, and the only person to win the Nobel Prize in two scientific fields. She was the first woman to become a professor at the University of Paris in 1906. In 1895 she married the French physicist Pierre Curie, and she shared the 1903 Nobel Prize in Physics with him and with the physicist Henri Becquerel for their pioneering work developing the theory of "radioactivity"—a term she coined. In 1906 Pierre Curie died in a Paris street accident. Marie won the 1911 Nobel Prize in Chemistry for her discovery of the elements polonium and radium, using techniques she invented for isolating radioactive isotopes.

Mark Watney (page 35): Fictional Astronaut stranded on Planet Mars, played by actor Matt Damon in the Movie "The Martian", book by Andy Weir.

Mark Twain (page 77): Pen name of Samuel Langhorne Clemens (1835 – 1910). An American writer, humorist, entrepreneur, publisher and lecturer. Best known by his pen name Mark Twain, He was lauded as the "greatest humorist the United States has produced," and William Faulkner called him "the father of American literature". His novels include The Adventures of Tom Sawyer (1876) and its sequel, the Adventures of Huckleberry Finn (1884).

Mars (pages 3,10): Planet Mars, the 4th planet from the Sun in our Solar System.

Marshall McLuhan (page 37): Herbert Marshall McLuhan (1911-1980) was a Canadian-born communications theorist and educator, whose work is among the cornerstones of the study of media theory. Author of the book, "This Spaceship Earth".

Marva Collins (page 11): 1936-2015. American educator best known for creating Westside Preparatory School, a private elementary school in the impoverished Garfield Park neighborhood of Chicago, IL which opened in 1975.

Matt Damon (page 35): Born October 8, 1970. Matthew Paige Damon is an American actor, producer, and screenwriter. Ranked among Forbes' most bankable stars, the films in which he has appeared have collectively earned over $3.12 billion at the North American box office, making him one of the highest-grossing actors of all time. He has received various awards and nominations, including an Academy Award and two Golden Globe Awards. Matt portrayed fictional astronaut Mark Watney in the movie The Martian directed by Ridley Scott, in which Mark Watney is stranded and forced to survive on planet Mars alone.

Maya Angelou (pages 7,62): Maya Angelou (1928-2014). An American poet, memoirist, and civil rights activist. She published seven autobiographies, three books of essays, several books of poetry, and is credited with a list of plays, movies, and television shows spanning over 50 years. She received dozens of awards and more than 50 honorary degrees. Angelou is best known for her series of seven autobiographies, which focus on her childhood and early adult experiences.

Mediocrity (page 67): Noun. The quality or state of being mediocre, meaning of only moderate quality or not particularly good.

Mercury (page 3): Planet Mercury, the 1st and closest planet to the Sun in our Solar System.

Michael Brown (pages 3,36): Born 1965. An American astronomer, professor of planetary astronomy at the California Institute of Technology (Caltech) since 2003. His team has discovered many Trans-Neptunian Objects (TNOs) including the "dwarf-planet" Eris which was originally thought to be bigger than Pluto, triggering a debate on the definition of a planet. Author of the book "How I Killed Pluto and Why It Had It Coming."

Microbes (page 41): Noun. A microorganism, especially a bacterium causing disease or fermentation.

Milky Way Galaxy (page 41): The Milky Way is the galaxy that contains our Solar System, with the name describing the galaxy's appearance from Earth: a hazy band of light seen in the night sky formed from stars that cannot be individually distinguished by the naked eye. The term Milky Way is a translation of the Latin "lacteal" and Greek. From Earth, the Milky Way appears as a band because its disk-shaped structure is viewed from within. Galileo Galilei first resolved the band of light into individual stars with his telescope in 1610. Until the early 1920s, most astronomers thought that the Milky Way contained all the stars in the Universe. Following the 1920 Great Debate between the astronomers Harlow Shapley and Heber Curtis, observations by Edwin Hubble showed that the Milky Way is just one of many galaxies.

Mons (page 20,56): From the Latin word for "mountain", Mons is a mountain on a celestial body. For example, the largest volcano on Mars is Olympic Mons.

Multispectral (page 62): Adjective. Operating in or involving several regions of the electromagnetic spectrum.

Murmurs of Earth (pages 92,93): A book by Carl Sagan and team about the record album carried aboard Voyagers 1 & 2.

Napoleon Hill (pages 13,59): Oliver Napoleon Hill (1883-1970). An American self-help author, best known for his 1937 book "Think and Grow Rich", based upon his interviews with Andrew Carnegie and 504 of America's wealthiest businessmen.

NASA (pages 29,44,47-57,60,65): National Aeronautics and Space Administration, a U.S. government agency responsible for science and technology related to air and space. See https://www.nasa.gov

National Air and Space Museum (page 5): Part of the Smithsonian Institution, a foundation for scientific research, established in 1836 and based in Washington, D.C. It operates more than a dozen museums and institutes in Washington and other cities. It originated with a bequest in the will of English chemist and mineralogist James Smithson (1765-1829). See https://airandspace.si.edu The Air and Space Museum is a museum in Washington D.C. established in 1946, dedicated to preserving the past and inspiring the future in Aviation, Astronomy, Exploration, Science and Engineering.

National Astronomical Observatory of Japan (30): The national center of astronomical research in Japan which owns and operates the Subaru Telescope on Mauna Kea, Hawaii.

Navigation (pages 21,22): Noun. The process of accurately determining one's position and planning and following a route.
Neil Armstrong (page 77): Neil Alden Armstrong (1930-2012). First person to walk on the Moon. American astronaut and aeronautical engineer, U.S. Naval Aviator, test pilot, university professor.
Neil deGrasse Tyson (pages 11,18): Born October 5, 1958. An American astrophysicist, planetary scientist, author, science communicator. Since 1996 he has been the director of the Hayden Planetarium at the Rose Center for Earth and Space in New York City. Author of the book "The Pluto Files".
Neptune (page 3,10): Planet Neptune, the 8th planet from the Sun, also a gas giant, in our Solar System.
New Horizons (pages 3,15,20,35,44-47,60,62,63,65): First spacecraft to visit and study "dwarf-planet" Pluto, the Pluto System and the Kuiper Belt.
Nightingale-Conant Corporation (page 9): An Illinois-based company founded by Earl Nightingale and Lloyd Conant which produces and sells professional audio learning programs on many topics. See www.nightingale.com
Nike (pages 11,18): Nike, Inc. is an American sportswear company headquartered in Beaverton OR, founded in 1964 as Blue Ribbon Sports by Bill Bowerman, a track and field coach at the University of Oregon and his former student Phil Knight. The company was renamed Nike, Inc. in 1978 and went public 2 years later.
Nix (page 30,48): A small Moon in the Pluto System.
Octave Chanute (page 6): 1832-1910. A French American civil engineer and aviation pioneer. He provided many budding aviation enthusiasts, including the Wright Brothers, with help and advice, and helped to publicize their flying experiments.
Oort Cloud (page 41): A spherical cloud of small rocky and icy astronomical bodies thought to orbit the sun beyond the orbit of Pluto and up to 1.5 light years from the sun and thought to be the source of comets. Its existence was proposed by J.H. Oort.
Oprah Winfrey (page 12): American talk show host, interviewer, television personality, former host and supervising producer of the top-rated award-winning The Oprah Winfrey Show, media leader, philanthropist.
Orville Wright (page 19): One of the two Wright Brothers, who invented and flew the 1st heavier than air powered airplane on December 17, 1903 at Kitty Hawk North Carolina.
Otto Lilienthal (page 6): 1848-1896. A German pioneer of aviation. The first person to make well-documented, repeated successful flights with gliders.
Out of the Darkness: The Planet Pluto (page 23): Book by Clyde Tombaugh and Patrick Moore, Foreword by Dr. James Christy.
Pale Blue Dot (page 16): A book and Television series about astronomy by astronomer Carl Sagan
Palomar Observatory (page 3): Established in 1928, an astronomical research observatory in San Diego County, CA located in the Palomar Mountain Range. Owned and operated by the California Institute of Technology (Caltech).
Parabolic (pages 64): Adjective. Relating to a parabola or having a the shape or curve of a parabola.
Patricia Edson (page 22): Maiden name of Clyde Tombaugh's wife.
Pearl Harbor (page 23): The attack on Pearl Harbor (Hawaii) was a surprise military strike by the imperial Japanese Navy Air Service upon the United States (a neutral country at the time) against the naval base at Pearl Harbor in Honolulu, just before 0800 on Sunday morning, December 7, 1941. The attack led to the United States' formal entry into World War II the next day.
Peggy Whitson (page 77): Retired American Astronaut, First Female Commander of the International Space Station during Expedition 16, biochemistry researcher, former NASA Chief Astronaut, Holds the record for the oldest woman spacewalker, record for the total spacewalks by a woman, oldest female astronaut ever in space, first female astronaut to command the International Space Station twice, record holder for total accrued time in space of 665 days---equivalent of a round trip to Mars, making her the most experienced astronaut to date).
PEPSSI (pages 62,63): Pluto Energetic Particle Spectrometer Science Investigation – a particle instrument aboard New Horizons spacecraft designed to capture Pluto's interaction with the solar wind.
Percival Lowell (pages 8,20,56,65): 1855-1916. An American businessman, author, mathematician, and astronomer. He founded Lowell Observatory in Flagstaff AZ and formed the beginning of the search for Planet X now named Pluto, which was discovered by Clyde Tombaugh 14 years after Lowell's death.
Persevere (page 13): Name of the newest 2021 Mars Rover. Noun, meaning steadfastness in doing something despite difficulty or delay in achieving success., Persistence, tenacity, determination, resolve, endurance.
Perseverance (pages 13, 24,32): Name of the SUV-sized Rover which NASA Jet Propulsion Laboratory (JPL) landed on Mars on February 18, 2021. It delivered the Ingenuity Mars Helicopter to the surface as well---the first powered aircraft to fly on another planet. The Rover successfully utilized the now-famous JPL Sky Crane Maneuver. See https://mars.nasa.gov/mars2020 and https://nasa.gov.perseverance.
Persistence (page 32): Noun. Firm continuance in a course of action despite difficulty or opposition.
Philosopher / Philosophy (pages 33,43,57): Philosopher: Noun. A person engaged or learned in philosophy, a thinker, theorist, dreamer, scholar. Philosophy: Noun. The study of the fundamental nature of knowledge, reality and existence, particularly as an academic study.
Phobos (page 18): The larger of Mars' two moons. In mythology, one of the two sons of Ares---the other is Deimos---the Greek counterpart of the Roman god of war named Mars. Phobos means fear or panic.
Planet (page 3): Noun. A word coined by the ancient Greeks, meaning "wanderer", but only clearly and scientifically defined recently by the International Astronomical Union in August 2006.
Planitia (pages 20,54,56): Latin word for plain, used in the naming g of plains on extraterrestrial planets, moons, dwarf-planet.

Pluto (page 3, 4,12-15, 29,30,47-57,60,65): Formerly the 9th planet in our Solar System. In August 2006, with the definition of the word "planet" defined scientifically, Pluto became a "dwarf-planet" and the First of a New Third Class of Solar System Objects: Trans-Neptunian Objects or Kuiper Belt Objects in our Solar System.

Pluto-ed – 2006 Word of the Year (page 3,23,38): The 2006 word of the year. Also Plutoed. Chosen by the American Dialect Society at its annual meeting. To Pluto is to demote, devalue someone or something, much like what happened to former planet Pluto in 2006 when the International Astronomical Union re-classified (Pluto-ed) Pluto as a "dwarf-planet".

Plutonium Dioxide (page 35): The radioactive fuel source used in Radioisotope Thermoelectric Generator (RTG). An RTG powers the New Horizons spacecraft.

Pluto Underground (page 32): A self-named team (in 1989) of scientists determined to develop a proposed mission to explore Pluto which would be accepted by NASA.

Powhatan (page 60): Powhatan or Virginia Algonquian is an extinct language belonging to the Eastern Algonquian subgroup of the Algonquian languages. It was spoken by the Powhatan people of tidewater Virginia and became extinct around the 1790s after its speakers were forced to switch to English. Kuiper Belt Object Arrokoth is named from the Powhatan language.

Profiles of the Future: An Inquiry Into the Limits of the Possible (page 53): Book by Arthur C. Clarke

Protractor (page 21): Noun. An instrument for measuring angles, typically in the form of a flat semicircle marked with degrees along the curved edge.

Queen (page 48): A British rock band formed in London in 1970 comprising Freddie Mercury, Brian May, Roger Taylor and John Deacon. Lead guitarist Brian May is also as Astrophysicist.

Radioactive/Radioactivity (page 35,41): RTG. A type of "nuclear battery" that uses an array of thermocouples to convert the heat released by the decay of radioactive material into electricity by the Seebeck effect. It has no moving parts. RTGs have been used as power sources in satellites and space probes and un-crewed remote facilities.

Radioisotope Thermoelectric Generator (RTG) (page 35):

RALPH (page 62): An instrument aboard the New Horizons spacecraft which serves as a color imager and an infrared mapping spectrometer.

Ralph Waldo Emerson (page 22): 1803-1882. An American essayist, lecturer, philosopher, abolitionist, poet, writer, who led the transcendentalist movement of the mid-19th century.

Red Shift (page 41): A displacement of the spectrum (light or radio waves) of an astronomical object toward the longer (Red) wavelengths, which indicates that the object is moving away from the observer. Like the Doppler Effect with sound waves.

Reflector/Reflecting Telescope (pages 4,52): A Reflecting Telescope or Reflector is a telescope that uses a single or a combination of curved mirrors that reflect light and form an image. This differs from a Refractor which uses lenses. The Reflector was invented by Sir Isaac Newton, hence it is called a Newtonian Reflector in his honor.

Refractor (pages 5,8): A lens-based telescope using lenses to magnify and focus light from distant objects. First documented use for aiming at the night sky by Galileo Galilei in the year 1610 in Italy.

Regio (page 20,56): In planetary geology, a large area of a planet or moon that is strongly differentiated in color or brightness. Example: Tombaugh Regio is the Heart Shaped region on Pluto.

Resilient (page 67): Adjective. Able to withstand or recover quickly from difficult conditions.

Resourceful (page 7): Adjective. Having the ability to find quick and clever ways to overcome difficulties.

REX (page 62,63): An instrument aboard the New Horizons spacecraft: Radio science Experiment, consisting of signal processing electronics combined with communications hardware on the main dish antenna.

Ridiculed (page 10): past tense of verb ridicule. Mocked, laughed at, made fun of.

Ridley Scott (page 35): Director of the movie "The Martian", starring actor Matt Damon as Mark Watney, stranded on Mars.

Robert Goddard (page 36, 78): American engineer, professor, physicist, inventor credited with creating & building world's 1st liquid-fueled rocket, which he launched on March 16, 1926. Holder of 214 patents. One of founders of modern rocketry.

Robert Harrington (pages 26,48): 1942-1993, an American Astronomer who worked at the United States Naval Observatory. Astronomer James Christy consulted with Robert after discovering bulges in the images of Pluto, which turned out to the Pluto's largest Moon. Some consider Rob to be co-discoverer of Charon.

Robert Ronstadt (page 9): Former Associate Professor of Management of Babson College where he served from 1975-1986, who developed the Entrepreneurship Program and the Corridor Principle. A recognized leader in technology commercialization, entrepreneurship and higher education. See **www.robertronstadt.com/about/my_bio**

Rob Stachie (page 32): Engineer at NASA Jet Propulsion Laboratory who took the 29-cent stamp of Pluto as a challenge to explore the tiny world.

Rock Band Styx (page 48): An American rock band from Chicago, IL which formed in 1972.

Roger Bannister (pages 51,73): Sir Roger Gilbert Bannister (1929-2018), a British middle-distance athlete and neurologist who ran the first sub-4-minute mile. He set this record on May 6, 1954 in Oxford: 3 minutes, 59.4 seconds. His record lasted 46 days

Roy E. Disney (page 58): The nephew of Walter Elias Disney of Walt Disney World fame.

ROYGBIV (page 41): An acronym for the sequence of hues commonly described as making up a rainbow: Red, Orange, Yellow, Green Blue, Indigo, and Violet; the order of colors seen in the visible light spectrum.

Sally Ride (pp. 23,78): 1951-2012. Dr. Sally Kristen Ride, American astronaut & physicist, 1st American woman in space.

Samuel Pierpoint Langley (page 6): 1834-1906. An American astronomer, physicist, inventor of the bolometer and an aviation pioneer. He also served as the 3rd Secretary of the Smithsonian Institution and was professor of astronomy at the University of Pittsburgh, and director of the Allegheny Observatory.

Saturn (pages 3,16): Planet Saturn, the 6th planet from the Sun, with the most spectacular ring system, and 2nd largest gas planet in our Solar System.

Saturn V (page 32): The Saturn V was an American super heavy-lifting launch vehicle certified for astronauts, and used by NASA between 1967 and 1973, developed for the Apollo lunar landing program, and later used to launch Skylab, the first American space station. The Saturn V was launched 13 times from the Kennedy Space Center in Florida.

Scientific Method (pages 38,41,42): Noun. A method of procedure that has characterized natural science since the 17th century, consisting of systematic observation, measurement, experiment, and the formulation, testing, and modification of hypotheses.

SCUBA (pages 28,41): Acronym for Self-Contained Underwater Breathing Apparatus. Improved by explorer Jacques Cousteau.

SDC (pages 20,62): Also see Student Dust Counter. Acronym for the Student Dust Counter instrument carried aboard New Horizons spacecraft.

Sedna (page 37): KBO 2003 UB12, discovered 11-14-2003 by Brown, Trujillo, Rabinowitz. About ¾ of Pluto's diameter.

Sequence (page 23): Noun. An order in which related events, movement, or things follow each other.

Shark Tank (page 15): An ABC television program on which ambitious entrepreneurs present their innovative business concepts to a team of experienced successful wealthy entrepreneurs ("sharks") in hopes of partnering with and receiving funding from the sharks. See Shark Tank Television on https://www.abc.com

Sherpa (page 56): One of the Tibetan ethnic groups native to the most mountainous regions of Nepal in the Himalayas.

Showalter, Mark (page 30): Senior Research Scientist, Planetary Astronomer with SETI institute, New Horizons team member, discoverer of Jupiter outer rings, initiated Hubble imaging of Pluto, discoverer of 2 Pluto Moons: Kerberos and Styx.

Sir Isaac Newton (pages 38,39,52): 1642-1726. An English mathematician, physicist, astronomer, theologian, and author who is widely recognized as one of the greatest mathematicians and most influential scientists of all time, and as a key figure in the scientific revolution. Newton invented the reflecting telescope (based upon a mirror), independently co-invented infinitesimal calculus, and discovered the laws of motion including the equation for gravity acting upon two masses at a distance.

SkyCrane Maneuver (pages 41, 56): A unique landing technique first developed by the Jet Propulsion Laboratory for the descent of NASA's Curiosity Rover to the Martian Surface on August 5, 2012 and was used again on February 18, 2021 for the Perseverance Rover landing on Mars. The maneuver involves stopping the descent about 20 meters above the Martian surface and lowering the rover slowly on cables until gentle touchdown---like lowering on a crane from the sky---then cutting the cables so the descent stage can fly away to a safe distance from the rover. See the very impressive YouTube video by JPL called "7 Minutes of Terror, describing the 7 minutes it takes the spacecraft with rover to enter Mars' Atmosphere, descend, and land. https://www.youtube.com/watch?v=GMoTcbzpK-o&t=8s

Smithsonian Institution (page 5): Also see National Air and Space Museum.

Southwest Research Institute (SwRI) (pages 44,49-57): Founded in 1947 by oil businessman Tom Slick and Headquartered in San Antonio, TX, the SwRI provides contract research and development services to government and industrial clients. One of the oldest and largest independent, nonprofit, applied research and development organization in the United States. **See** https://www.swri.org

SpaceShipOne (page 65): An experimental air-launched rocket-powered aircraft with sub-orbital spaceflight capability. A spaceplane designed to carry 3 people in a sea-level-pressurized cabin, which completed the 1st crewed private spaceflight in 2004. It's launch aircraft was named White Knight. Both crafts were developed and flown by Mojave Aerospace Ventures, a joint venture between Paul Allen and Scaled Composites, an aviation company owned by Burt Rutan. SpaceShipOne now hands in the National Air and Space Museum in Washington, D.C.

Speed of Light (page 63): In a vacuum (i.e., space), the speed of light, denoted by "c" is a universal physical constant important in many areas of physics. It is literally the speed which light photons travel in a vacuum. Its exact value is defined as 299,792,458 meters per second or approximately 300,000 kilometers per second or 186,000 miles per second. By international agreement, a meter is defined as the length of the path travelled by light in vacuum during time interval of 1/299792458 sec.

Sputnik (page 56): The 1st artificial Earth satellite launched into an elliptical low Earth orbit by the USSR on October 4, 1957 as part of the Soviet space program.

Star Trek (page 97): An American science-fiction television series and media franchise created by Gene Roddenberry that follows the adventures of the starship USS Enterprise and her crew. Trademarks owned by CBS Studios, Inc.

Stacy Winstein (page 32): A Jet Propulsion Laboratory, California Institute of Technology, Solar System Exploration Department Member, Deputy payload manager, interplanetary mission design and astrodynamics project and systems engineering and technical manager. Key player in taking on the 29-cent stamp challenge in pushing for the exploration of Pluto.

Styx (pages 27,30,48): Rock Band Styx: An American Rock Band of the early 1970's and 1980s, which formed in Chicago in 1972, continued reuniting in 1990. Pluto's Moon Styx: A small natural satellite of Pluto whose discovery was made by a team led by astronomer Mark Showalter using 14 sets of images taken by the Hubble Space Telescope. Discovery was announced 7-11-12.

Subaru Telescope (page 30): An 8.2 meter optical-infrared telescope at the summit of Mauna Kea on the Big Island of Hawaii, operated by the National Astronomical Observatory of Japan. See https://subarutelescope.org/en

Success (page 79): The progressive realization of a worthy goal of your choosing. Earl Nightingale's definition.

Success Mastery Academy (pages 43, 50): Audio Learning Program/Seminar about business success by Brian Tracy.

SWAP (page 63): Solar Wind Around Pluto (SWAP) science experiment carried aboard the New Horizons spacecraft to Pluto.

Sylvia Kuiper (page 45): Daughter of Dutch Astronomer & planetary scientist, author and Professor Gerard Peter Kuiper, for who the Kuiper Belt is named.

Syndicated Columnist (page 63): A writer who produces regular short articles typically on a specific theme or subject and sells them to a service that distributes them for her or him, usually over wide geographic regions.

Tedium / Tedious (page 15): Noun/Adjective. The state of being tedious, dullness, boredom, sameness, unchanging.

Telegram (page 19): Noun. A message sent by telegraph and then delivered in written or printed form.

Telescope (page 5,10): Noun. An optical instrument designed to make distant objects appear nearer, containing an arrangement of lenses or curved mirrors and lenses to focus the light. A radio telescope is a non-optical antenna and radio receiver used to detect radio waves from astronomical objects, based upon a parabolic shape.

Tenacity (page 32): Noun. The quality or fact of being incredibly determined, determination, persistence.

Tenzing Norgay (pages 51,73): Tenzing Norgay GM OSN, born Namgyal Wangdi, and referred to as Sherpa Tenzing, was a Nepali-Indian Sherpa mountaineer. He was one of the first two individuals known to reach the summit of Mount Everest, which he accomplished with Edmund Hillary on 29 May 1953.

Terra (page 56): Noun. In Geographic astronomy, an alternate name for Planet Earth, Latin name for the planet, extensive land masses found on various solar system bodies.

The Case for Pluto: How a Little Planet Made a Big Difference (pages 15,19,34): Book by author Alan Boyle

The Correspondence of Isaac Newton (page 52): Considered as a single group of letters written by Isaac Newton (from 1676 – 1687, and first published by the Royal Society), the Newton-Cotes correspondence is the largest and most important section of Isaac Newton's scientific correspondence that we have. It is available in 7 paperback volumes from www.Amazon.com

The Martian (page 35): Book by author Andy Weir, made into an excellent movie starring Matt Damon as stranded-on-Mars-astronaut Mark Watney, directed by Ridley Scott.

Theorized (page 10): Verb. Past tense of Theorize. To form a theory or theories about some subject. A Theory is a supposition or system of ideas intended to explain something, especially one based upon general principles independent of the thing to be explained, and which is subject to independent and open investigation, observation, challenge, and objective observations and evidence. Also: hypothesis, thesis, proposition.

The Pluto Files (pages 18, 37): Book by astronomer Neil deGrasse Tyson.

Think and Grow Rich (page 59): 1937 Book of 17 Success Principles summarizing the success philosophies of 504 of the most successful American businessmen, as commissioned by Andrew Carnegie in 1908, authored by Napoleon Hill.

Tholins (pages 48,60): red organic compounds formed when hydrocarbon molecules are irradiated by sunlight or cosmic rays.

Thomas Edison (page 34,42,43,66): Thomas Alva Edison (1847-1931). An American Inventor and businessman, described as America's greatest inventor, awarded 1093 U.S. Patents and many more global patents for his inventions, including the phonograph, motion picture camera, incandescent electric light and many more inventions.

Trans Neptunian Object (TNO) (page 3, 36): TNO. Classification given to any minor planet in the solar system that orbits the sun at a greater average distance than Neptune, within the Kuiper Belt. Pluto was the 1st TNO discovered in 1930.

Ultima Thule (page 60): Name first given to KBO discovered by Marc Buie on June 26, 2014, renamed Arrokoth, visited by New Horizons spacecraft on January 1, 2019---the farthest Kuiper Belt Object visited by a spacecraft after its Pluto System flyby.

University of Kansas (page 22): The University of Kansas is a public research university with its main campus in Lawrence, Kansas, and several satellite campuses, research and educational centers, medical centers, and classes across the state of Kansas. See https://www.ku.edu

Uranus (page 3): Planet Uranus, the 7th planet from the Sun, also a gas giant, discovered by William Herschel.

U.S. Army Corps of Engineers (page 55): An Engineering formation of the Unit States Army that has 3 primary missions: Engineering, Military Construction, Civil Works.

U.S. Department of Energy (page 62): The United States Department of Energy (DOE) is a cabinet -level department of the United States Government concerned with the United States ' policies regarding energy and safety in handling nuclear material.

U.S. Naval Observatory (pages 26,27): The United States Naval Observatory (USNO) is one of the oldest scientific agencies in the United States, with a primary mission to produce Positioning, Navigation and Timing (PNT) for the United States Navy and the United States Department of Defense.

U.S.S. Enterprise (page 97): The name given to many vessels throughout history, including an 18th century British Supply Sloop, first U.S. nuclear-powered aircraft carrier (now decommissioned CVN-65), current United States Navy Aircraft Carrier (CVN-80), flight test Space Shuttle, and future Starship from the Star Trek media franchise (trademarks owned by CBS Studios, Inc.).

Varuna (page 37): KBO 2000 WR106 discovered on 11-28-2000.

Venus (page 3): Planet Venus, the 2nd planet (and hottest) from the Sun in our Solar System.

Venera (pages 20,56): The Venera program was the name given to a series of space probes developed by the Soviet Union between 1961 and 1984 to gather information about the planet Venus. Ten probes successfully landed on the surface of the planet, including the two Vega program and Venera-Halley probes, while thirteen probes successfully entered Venus' atmosphere. Due to the extremely high temperatures and pressures at the surface of Venus, the probes only survived from 23 minutes to two hours. Features on Pluto were named in honor of the Venera program.

Venetia Burney Dust Counter (page 63): Also see Student Dust Counter or SDC, a science instrument carried aboard the New Horizons Spacecraft to Pluto and beyond into the Kuiper Belt.

Venetia Burney (Phair) (pages 18, 20, 56,65): Venetia Katharine Douglas Burney (1918-2009), married name Phair. As an English girl of 11 years old, she was credited by Clyde Tombaugh with first suggesting the name Pluto for the planet he discovered in 1930. She was living in Oxford, England, at the time. As an adult she worked as an accountant and a teacher of economics & math in England. See https://solarsystem.nasa.gov/people/2902/venetia-burney-phair-1918-2009

Vesto (V.M.) Slipher (pages 8,12,17): Vesto Melvin Slipher (1875-1969). An American Astronomer who performed the first measurements of radial velocities for galaxies. He was the first to discover that distant galaxies are redshifted, thus providing the first empirical basis for the expansion of the universe. While working at the Lowell Observatory in Flagstaff, AZ, Vesto hired Clyde Tombaugh to search for Planet X, which Clyde discovered on February 18, 1930. Planet X was later named Pluto.

Voyager 1 & 2 (page 16,19,56,92,97): Twin unmanned spacecraft launched from Earth in 1977 with a mission to explore Jupiter and Saturn, and beyond to the outer gas giant planets Uranus and Neptune. See https://www.voyager.jpl.nasa.gov and https://spaceplace.nasa.gov/voyager-to-planets

Walt Disney (pages 20,58): American entrepreneur, animator, writer, voice actor, film producer, television host, pioneer of the American animation industry, founder of Disney Brothers Studio, creator of Mickey Mouse in 1928, founder of Disneyland in 1955, The Mickey Mouse Club, Walt Disney World in 1971.

Walt Disney World (page 58): Theme park created by Walt Disney, in Orlando FL. See https://disneyworld.disney.go.com

Wernher von Braun (pages 32,34,55,79): German-born American Aerospace Engineering and space architect, rocket scientist. The Saturn V vehicle was developed by the Marshall Space Flight Center in Huntsville, AL under the direction of von Braun. The Saturn V rocket carried all the Apollo astronauts to the Moon from July 1969 (Apollo 11) – December 1972 (Apollo 17).

White Sands Missile Range (page 23): White Sands Missile Range is a military testing area operated by the United States Army. The range was originally established as the White Sands Proving Ground on July 9, 1945. White Sands National Park is located within the range.

Wikipedia (page 56): A version of the encyclopedia Wikipedia, an open-content online encyclopedia, collaboratively developed over the World Wide Web.

William Herschel (page 15): Frederick William Herschel (1738-1822), a German-born British astronomer, composer and brother of fellow astronomer Caroline Herschel, with whom he worked. He built his first large telescope in 1774 and spent 9 years observing and cataloguing double-stars and nebulae. He published a catalog in 1802 with 2500 objects and 1820 with 5000 objects. On March 13, 1781, he discovered a new object. After several weeks of confirmation, it was named Planet Uranus.

Sir Winston Churchill (pages 7, 34): Sir Winston Churchill, British Statesman, orator, author, Former British Prime Minister from 1940-1945, 1951-1955, he rallied the British people during World War II, leading his country from the brink of defeat to victory.

Wilbur and Orville Wright, The Wright Brothers (pages 5,6,20,24,56): The Wright Brothers invented and flew the 1st heavier than air powered airplane on December 17, 1903 at Kitty Hawk North Carolina. They owned a bicycle shop in Dayton OH, now a museum site---go visit it!

Xena (pages 3, 37): Name first given to KBO 2003 UB313, later Eris, discovered 1-5-2005, 1st thought to be larger than Pluto.

X-Rays (page 41): an electromagnetic wave of high energy and short wavelength, which can pass through many materials opaque to light, widely used in medical and dental diagnoses to "see through" the body to image bones and teeth.

Yanni (page 59): Yiannis Chryssomallis (born in 1954), known professionally as Yanni, is a Greek American composer, keyboardist, pianist, and music producer.

Yoda (page 11,79): Fictional character & Jedi Master in the Star Wars franchise (now owned by Disney), first appearing in the 1980 George Lucas film 'The Empire Strikes Back'.

Zig Ziglar (page 61,79): 1926-2012, American author, businessman, sales professional, motivational speaker.

Replica of the plaque attached to the leg of the Apollo 11 Lunar Module, waiting for future visitors to read. (Credit: NASA and the Author)
The Voyager Records (on right) are "cosmic messages in a bottle"!
See Page 71 for Recorded Greetings, including Nick Sagan, at age 6.

Cosmic Map & Playing Instructions covering the "Record" aboard Voyagers 1 & 2, launched in 1977, containing 118 photos, greetings in 55 Languages, including whale & 90 min of music! (Credit: NASA, see book: "Murmurs of Earth" By Dr. Carl Sagan). What time capsule message or artifacts would you send into the galaxy?

Dual Bibliography: Clyde, Pluto, Astronomy (AND Success!)
Publications to Learn More About Clyde Tombaugh, Discoverer of Pluto, Pluto, Astronomy, Wright Brothers, Aviation, The Moon, New Horizons Mission:

- Aguilar, David. *13 Planets: The Latest View of the Solar System.* Washington, D.C.: National Geo., 2011.
- Boyle, Alan. *The Case for Pluto.* Hoboken, NJ: John Wiley & Sons, Inc., 2010.
- Brown, Mike. *How I Killed Pluto and Why It Had It Coming.* New York, NY: Spiegel & Grau, 2010.
- Caruso, Kevin. *Back to the Moon: Mankind Returns to the Lunar Surface.* Streamwood, IL: Aerospace1 Publications, 2001. (ISBN# 0-9705150-0-6)
- Cernan, Eugene, and Don Davis. *The Last Man on the Moon.* New York, NY: St. Martin's Press, 1999.
- Chaikin, Andrew. *A Man on the Moon: The Voyages of the Apollo Astronauts.* New York, NY: Viking Press, 1994.
- Daniels, Patricia and Robert Burnham. *The New Solar System: Icy Worlds, Moons, and Planets Redefined.* Washington, D.C.: National Geographic, 2010.
- Houk, Rose. *From the Hill: The Story of Lowell Observatory.* Salt Lake City, UT: Paragon Press, 2017.
- International Astronomical Union. *IAU 2006 General Assembly Result of the IAU Resolution votes.* Prague: IAU0603 Press Release, 2006. https://www.iau.org/news/pressreleases/detail/iau0603/
- Jones, Barrie. *Pluto: Sentinel of the Outer Solar System.* New York, NY: Cambridge University Press, 2010.
- Kortenkamp, Steve. *Why Isn't Pluto a Planet?* Mankato, MN: Capstone Press, 2007.
- Landau, Elaine. *Pluto: From Planet to Dwarf.* New York, NY: Children's Press, 2008.
- Levy, David. *Clyde Tombaugh: Discoverer of Planet Pluto.* Cambridge, MA: Sky Publishing Corp., 2006.
- Lew, Kristi. *The Dwarf Planet Pluto.* New York, NY: Marshall Cavendish, 2010.
- Marschall, Laurence, and Stephen Maran. *Pluto Confidential: An Insider Account of the Ongoing Battles over the Status of Pluto.* Dallas, TX: Benbella Books, Inc., 2009.
- Minard, Anne. *Pluto and Beyond: A Story of Discovery, Adversity, and Ongoing Exploration.* Flagstaff, AZ: Northland Publishing, 2007.
- Nardo, Don. *Pluto.* Farmington Hills, MI: KidHaven Press, 2003.
- NASA. *Solar System Exploration: Pluto.* NASA, 2022. https://solarsystem.nasa.gov/planets/dwarf-planets/pluto/overview/
- Owens, L.L.. *Pluto and Other Dwarf Planets.* Mankato, MN: The Child's World, 2011.
- Portman, Michael. *Why Isn't Pluto a Planet?* New York, NY: Gareth Stevens Publishing, 2013.
- Rusch, Elizabeth and Guy Francis. *The Planet Hunter: The Story Behind What Happened to Pluto.* Flagstaff, AZ: Rising Moon, 2007.
- Sagan, Carl. *Cosmos.* New York, NY: Random House, Inc., 1980.
- Sagan, Carl, et al. *Murmurs of Earth.* New York, NY: Random House, 1978.
- Schilling, Govert. *The Hunt for Planet X: New Worlds and the Fate of Pluto.* New York, NY: Copernicus Books, 2009.
- Schindler, Kevin and Will Grundy. *Pluto and Lowell Observatory.* Charleston, SC: The History Press, 2018.
- Scott, Elaine. *When Is a Planet Not a Planet?: The Story of Pluto.* New York, NY: Clarion Books, 2007.
- Stern, Alan, and David Grinspoon. *Chasing New Horizons: Inside the Epic First Mission to Pluto.* New York, NY: Picador, 2018.
- Stern, Alan, and Jacqueline Mitton. *Pluto and Charon: Ice Worlds on the Ragged Edge of the Solar System.* New York, NY: Wiley-VCH, 2005.
- Sutherland, Paul. *Where Did Pluto Go?: A Beginner's Guide to Understanding the "New Solar System".* East Sussex, United Kingdom: Ivy Press Limited, 2009.
- Tombaugh, Clyde, and Patrick Moore. *Out of the Darkness: The Planet Pluto.* Harrisburg, PA: Stackpole Books, 1980.
- Tyson, Neil deGrasse. *The Pluto Files: The Rise and Fall of America's Favorite Planet.* New York, NY: W.W. Norton & Company, Inc., 2009.
- Wetterer, Margaret, and Laurie Caple. *Clyde Tombaugh and the Search for Planet X.* Minneapolis, MN: Carolrhoda Books, Inc., 1996.
- Weintraub, David. *Is Pluto a Planet?: A Historical Journey through the Solar System.* Princeton, NJ: Princeton University Press, 2007.

- Weir, Andy. *The Martian.* New York, NY: Broadway Books, 2011.
- Wright, Orville. *How We Made The First Flight.* Washington, D.C.: Federal Aviation Administration (FAA) Publication APA-6-155-88, 1986. https://faa.gov/education/educators/curriculum/k12/media/k-12_how_we_made_the_first_flight_orville_wright.pdf

Publications and Professional Audio Learning Programs & Classics about Life-Lessons (Kuiper Belt Wisdom), Philosophies for Successful Living and Biographies:

- Adkins, Jan. *Thomas Edison Biography.* New York, NY: DK Publishing, 2009.
- Axelrod, Alan. *Edison on Innovation: 102 Lessons in Creativity for Business and Beyond.* Hoboken, NJ: Jossey-Bass, 2008.
- Barnes, Jeffrey. *The Wisdom of Walt: Leadership Lessons from the Happiest Place on Earth.* Lake Placid, NY: Aviva Publishing, 2015.
- Blair, Gary Ryan. *Everything Counts!* Hoboken, NJ: John Wiley & Sons, Inc., 2010.
- Branson, Richard. *Screw It, Let's Do It.* Croydon, Great Britain: Virgin Books, 2009.
- Brown, Les. *Live Your Dreams.* New York, NY: Avon Books, 1992.
- Conwell, Russell. *Acres of Diamonds.* Old Tappan, NJ: Fleming H. Revell Company, 1988.
- Covey, Stephen. *The 7 Habits of Highly Effective People.* New York, NY: Simon & Schuster, Inc., 1989.
- Covey, Stephen. *The 8th Habit: From Effectiveness to Greatness.* New York, NY: Sound Ideas, 2004. (CDs)
- Cummuta, John. *Transforming Debt into Wealth.* Wheeling, IL: Nightingale-Conant, 2010. (CDs)
- Dry, Sarah. *Curie.* London, Great Britain: Haus Publishing Limited, 2000.
- Gitomer, Jeffrey. *Jeffrey Gitomer's Little Gold Book of YES! Attitude: How to Find, Build, and Keep a YES! Attitude for a Lifetime of SUCCESS.* Upper Saddle River, NJ: Financial Times Press, 2007.
- Hill, Napoleon. *The Law of Success.* New York, NY: Penguin Group, 2008. (original publication date 1928)
- Hill, Napoleon. *The Science of Personal Achievement: The 17 Principles of Success.* Wheeling, IL: Nightingale-Conant Corporation, 1992 (CDs).
- Hill, Napoleon. *Think and Grow Rich.* New York, NY: Ballantine Books, 1988 (16th printing).
- Kotte, John. *A Sense of Urgency.* Boston, MA: Harvard Business Review Press, 2008.
- Krensky, Stephen. *Benjamin Franklin.* New York, NY: DK Publishing, 2008.
- Mason, John. *You're Born An Original, Don't Die A Copy!* Altamonte Springs, FL: Insight Intrnat'l, 1993.
- Nightingale, Earl. *Lead the Field.* Niles, IL: Nightingale-Conant Corporation, 1986 (CDs).
- Nightingale, Earl. *The Essence of Success.* Niles, IL: Nightingale-Conant Corporation, 1993.
- Parker, Sam, and Mac Anderson. *212 Degrees: the extra degree.* Naperville, IL: Simple Truths, 2006.
- Peters, Tom. *The Brand You 50.* New York, NY: Random House, Inc., 1999.
- Peters, Tom. *The Pursuit of Wow!.* New York, NY: Vintage Books, 1994.
- Reeve, Christopher. *Nothing is Impossible.* New York, NY: Random House, Inc., 2002.
- Rohn, Jim. *The Seasons of Life.* Irving, TX: Jim Rohn International, 1996.
- Rohn, Jim. *7 Strategies for Wealth & Happiness: Power Ideas from America's Foremost Business Philosopher.* Rocklin, CA: Prima Publishing, 1996.
- Rohn, Jim. *Jim Rohn's Weekend Leadership Event.* Southlake, TX: Jim Rohn International, 2004 (DVDs).
- Rohn, Jim. *The Five Major Pieces to the Life Puzzle.* Dallas, TX: Great Impressions Printg/Grphs, 1991.
- Sullenberger, Chesley (Sully) III, and Douglas Century. *Making a Difference.* New York, NY: HarperCollins Publishers, 2012.
- Tracy, Brian. *The New Psychology of Achievement.* Niles, IL: Nightingale-Conant Corporation, 2008 (CDs).
- Tracy, Brian. *Success Mastery Academy.* Southlake, TX: YourSuccessStore.com, 2001 (CDs).
- Tracy, Brian. *Many Miles to Go: A Modern Parable for Business.* Irvine, CA: Entrepreneur Media, Inc., 2003.
- Tracy, Brian. *The Universal Laws of Success and Achievement.* Wheeling, IL: Nightingale-Conant Corporation, 1992 (CDs).
- Waitley, Denis. *The Psychology of Winning.* New York, NY: Berkley Books, 1984.
- Waitley, Denis. *The Winner's Edge.* New York, NY: Berkley Books, 1983.
- Woodside, Martin. *Thomas Edison: The Man Who Lit Up the World.* New York, NY: Sterling Publishing Company, Inc., 2007.
- Yager, Dexter, Sr. *Don't Let Anyone Steal Your Dream.* Charlotte, NC: Internet Services Corporation, 1978.
- Ziglar, Zig. *See You at the Top.* Gretna, LA: Pelican Publishing Company, Inc., 1989 (45th printing).

Awesome Web Sites for Further Exploration and Ongoing Education about Clyde, Pluto, Astronomy AND Philosophies for Successful Living!

As with everything else in the Universe, Websites are in constant motion. These were current at time of writing and serve as launching points! **Study, Learn, Prepare yourself in whatever subjects you choose!**

- Academy of Achievement & Clyde Tombaugh https://achievement.org/achiever/clyde-tombaugh/
- Alan Stern https://en.wikipedia.org/wiki/Alan_Stern Thank you Dr. Stern!
- Alden Tombaugh, Annette Tombaugh, Gerard Kuiper, Sylvia Kuiper p.38 The children of Pluto: reflections on Clyde Tombaugh – Astronomy Now
- American Booksellers Association https://www.bookweb.org
- American Dialect Society https://americandialect.org
- Andy Weir, Author of The Martian https://www.andyweirauthor.com
- Ann Landers www.annlanders.com
- Apollo Lunar Surface Journal. **Wow! This is an amazing and complete transcript and photo archive** of the six Apollo (11-17) Lunar Surface Mission Explorations of the Moon from 1969-1972. **Check it out!** https://www.hq.nasa.gov/alsj
- Arizona Memory Project (Pluto Telegram) part of Arizona State Library, Archives and Public Records, Division of AZ Secretary of State https://azmemory.azlibrary.gov/digital/collection/loaselect/id/15/rec/52
- Arthur C. Clarke Arthur C. Clarke - Wikipedia
- Astronomy Picture Of the Day https://apod.nasa.gov
- Babson College http://www.babson.edu
- Bodleian Library https://www.bodleian.ox.ac.uk
- Brian Tracy https://www.briantracy.com Thank you Brian for decades of practical success philosophies!
- Carl Sagan https://carlsagan.com
- Clyde Tombaugh, Academy of Achievement https://achievement.org/achiever/clyde-tombaugh/
- Corridor Principal, Dr. Robert Ronstadt, Babson College https://thebusinessprofessor.com/lesson/corridor-principle-definition/
- Daily Telegraph https://www.dailytelegraph.com.au/news/ english-girl-venetia-burney-11-who-suggested-the-name-pluto-in-1930-remains-the-only-female-to-have-named-a-planet/news-story/a4f4fb88f809ef7c3b983beefb21ad51
- Deep Space Network https://eyes.nasa.gov/dsn
- Denis Waitley www.deniswaitley.com Thank you Denis for your Psychology of Winning!
- Disney https://www.disneyworld.disney.go.com
- Eris ("dwarf-planet") https://solarsystem.nasa.gov/dwarf-planets/eris/in-depth/
- Federal Aviation Administration https://faa.gov
- Franklin Institute & Thomas Edison https://www.fi.edu/history-resources/edisons-lightbulb
- Gemini Observatories, North and South, Hawaii and Chile. https://www.gemini.edu and Check out the spectacular panorama of the Mauna Kea Observatories in Hawaii by Frank Ravizza at https://en.wikipedia.org/wiki/Mauna_Kea_Observatories#/media/File:Panorama_of_Mauna_Kea_Observatories.jpg Thank you Gemini Observatory for years of astronomical learning & memories at Mauna Kea!
- Hubble Space Telescope www.hubblesite.org
- International Astronomical Union, Pluto vote https://www.iau.org/news/pressreleases/detail/iau0603/
- Jacques Cousteau, see www.cousteau.com or www.cousteau.org
- James Christy https://www.britannica.com/biography/James-W-Christy
- Dr. Jeff Goldstein, Astrophysicist, Founder of Journey Through The Universe, National Center for Earth and Space Science Education. Thank you Jeff for inspiring & introducing me to Journey Through The Universe! https://www.journeythroughuniverse.org/downloads/Researchers/pdt_Goldstein.pdf
- Jeffrey Gitomer (Yes attitude) www.gitomer.com Thank you Jeff for your "out of the box" thinking!
- Jet Propulsion Laboratory, California Institute of Technology, The Planets https://www.jpl.nasa.gov

and check out the amazing Mars Rover Entry, Descent & Landing Video (including the creative JPL SkyCrane Maneuver) at https://www.youtube.com/watch?v=M4tdMR5HLtg Thanks for inspiring, JPL!
- Jim Rohn www.jimrohn.com Thank you Jim for your practical positive forward philosophy & action!
- John F. Kennedy Presidential Library www.jfklibrary.org
- Johns Hopkins University Applied Physics Laboratory https://www.jhuapl.edu
- Journey Through The Universe (JTTU) Week at Gemini Observatory, Hawaii. Also see Dr. Jeff Goldstein. https://www.gemini.edu/node/11817 and JTTU video: https://youtu.be/iC5JOaDGEg4 (author presenting scale models, Newton's 3rd Law of Action & Reaction with indoor-safe rockets, "World's Cheapest Space Suit" at 0:22, 0:28, 0:55, 1:26, and 1:38 times) Thank you Journey Team for amazing Journey memories!
- Lowell Observatory/Lowell Observatory Archives https://lowell.edu Thank you Lauren Amundson!
- Malin Space Science Systems, Camera Systems for Solar System Explorations www.msss.com
- Marie Curie www.britannica.com/biography/Marie-Curie
- Menlo Park Museum (Edison's Lab) https://www.menloparkmuseum.org/history
- Napoleon Hill (Success Philosophy) https://www.naphill.org and www.nightingale.com
- NASA https://www.NASA.gov and all topics (A-Z) https://www.NASA.gov/topics Thank you NASA!
- NASA Solar System Exploration https://solarsystem.nasa.gov/planets/dwarf-planets/pluto/overview/
- New Horizons Mission and Spacecraft Science, www.pluto.jhuapl.edu/Mission/Spacecraft.php
- Nightingale-Conant Corporation, Ongoing Professional Education www.nightingale.com Thank you NC!
- Percival Lowell https://en.wikipedia.org/wiki/Percival_Lowell
- Perseverance Rover on Mars https://mars.nasa.gov/mars2020 and https://nasa.gov.perseverance.
- Pluto Facts https://nssdc.gsfc.nasa.gov/planetary/factsheet/plutofact.html
- Pluto for Kids https://spaceplace.nasa.gov/ice-dwarf/en
- Pluto Home Page https://nssdc.gsfc.nasa.gov/planetary/planets/plutopage.html
- Pluto Mission & New Horizons Spacecraft https://pluto.jhuapl.edu/Mission/Spacecraft.php
- Queen Lead Guitarist Brian May https://brianmay.com
- Robert Harrington https://solarsystem.nasa.gov/moons/pluto-moons/charon/in-depth
- Rock Band Styx Styx (band) - Wikipedia
- Rare Newspapers https://www.rarenewspapers.com/view/559832 and /588339
- Roger Bannister https://en.wikipedia.org/wiki/Roger_Bannister
- Scientific Method, Wikipedia 1200px-The_Scientific_Method_as_an_Ongoing_Process.svg.png (1200×988) (wikimedia.org)
- Smithsonian Institution, National Air and Space Museum https://www.si.edu
- Solar System Exploration, WOW! https://solarsystem.nasa.gov and www.explanet.info
- Southwest Research Institute https://www.swri.org
- Space News Stories for the Week https://www.space.com
- Space Math & Fractions https://spacemath.gsfc.nasa.gov
- Success Mastery Academy www.BrianTracy.com Thank you Brian for decades of practical wisdom!
- Sun, daily near-real-time images https://sohowww.nascom.nasa.gov
And https://www.nasa.gov/mission-pages/sdo/the-sun-now/index.html
- Sir Edmund Hillary - Sir Edmund Hillary and Tenzing Norgay - 1953 Everest (nationalgeographic.com)
- Sir Isaac Newton https://www.britannica.com/biography/Isaac-Newton
- Subaru Telescope https://subarutelescope.org/en Thank you Subaru for the Tour atop Mauna Kea!
- Tenzing Norgay https://en.wikipedia.org/wiki/Tenzing_Norgay
- Tom Peters www.tompeters.com Thank you Tom for sharing your excellence wake-up philosophy!
- University of Kansas https://www.ku.edu
- U.S. Naval Observatory https://www.usno.navy.mil/USNO
- Venetia Burney https://en.wikipedia.org/wiki/Venetia_Burney
- Voyagers 1 & 2 https://spaceplace.nasa.gov/voyager-to-planets and https://www.voyager.jpl.nasa.gov
- Wernher von Braun https://www.nasa.gov/centers/marshall/history/vonbraun/bio.html
- Wright Brothers – National Park Service. Wright Brothers (U.S. National Park Service) (nps.gov) and https://wright.nasa.gov
- Zig Ziglar www.ziglar.com Thank you for your wisdom, positive philosophies and living what you taught!

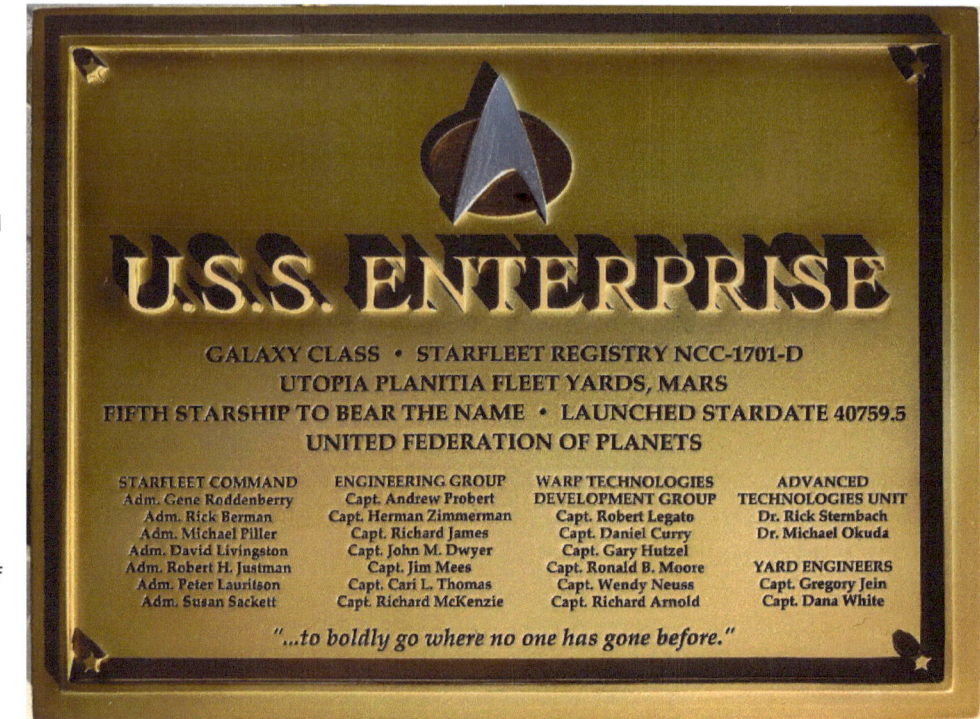

(Photo Credit: The Author, Replica of the Dedication Plaque aboard the science fiction Starship *Enterprise* NCC-1701-D from *Star Trek, The Next Generation*, television series, logo, Trademarks of CBS Studios, Inc.)

Twenty-fourth-century history will record that in the year 2363 (Stardate 40759.5), a Galaxy-Class Starship named <u>U.S.S. Enterprise</u> (Starfleet Registry Number NCC-1701-D) will be launched from the Utopia Planitia Fleet Yards in Mars' orbit.

This will be the 5th starship (641 meters in length) and one of many vessels throughout history to bear the name *Enterprise,* including:

- an 18th century British supply sloop
- the 1st nuclear-powered aircraft carrier in the United States Navy (Hull # CVN-65)
- the 1st NASA Space Shuttle Orbiter atmospheric test flight vehicle
- the newest Gerald R. Ford-Class Aircraft Carrier presently being built by the United States Navy (USS Enterprise, Hull # CVN-80---the 9th U.S. Naval Vessel to bear the name)

On the bridge of that future Starship will be a dedication plaque which charges the ship with its mission:

"…to boldly go where no one has gone before."

Throughout history, such plaques have represented a distinctly human element aboard vessels designed to explore. They have nothing to do with the function of the vessel. **They represent the mission of the ship, the spirit of the captain, crew, craft designers. This is why** we name our boats and spacecraft, why we attached plaques to Lunar Modules, flew messages to deep space with Pioneers & Voyagers & sent spacecraft like New Horizons carrying a 29-cent postage stamp & the names of 434,000 people who supported the mission. **We are curious creatures who must explore, learn and grow!**

In this spirit, I wish for you this same mission for your life: Boldly Go!

Author Biography & Gratitude:

Kevin Caruso works as an American Society for Quality (ASQ) Certified Manager of Quality & Organizational Excellence and Supplier Quality Engineer for a leading global aerospace company. He has worked in industrial, defense, and aerospace electronics manufacturing for 30 years.

He's an electrical engineer, private pilot and proud father of two wonderful young adults. He authored and published the middle-school Lunar Science Book **"Back To The Moon"**, in 2001.

Kevin has served students and teachers for **29 years as a NASA Aerospace Education Volunteer** and former **9-year NASA JPL Solar System Ambassador Volunteer for Illinois,** sharing **Apollo Moon Rocks** and **"The World's Cheapest Space Suit"** with school classrooms, libraries, science centers & senior centers throughout the country.

Kevin Caruso near the 13,796-foot (4205 meters) Hawaiian Sacred Summit of Mauna Kea on the Big Island of Hawaii, an 11,228-acre Science Reserve hosting 13 of the world's largest and most advanced telescopes. It looks just like the surface of Mars, including the snow! Yes, snow in Hawaii! How cool is that? (Credit: Author)

Kevin's life-long passion for astronomy began when he peered at Jupiter through a friend's telescope in 6th grade. **His favorite 10-year volunteer activity includes sharing the excitement of astronomy, engineering & space exploration** as a visiting-Engineer, Amateur Astronomer & Space Author with the students and teachers on the **Big Island of Hawaii** as part of Hilo's annual *"Journey Through The Universe" week.* During each "Journey Week" (still today), Mauna Kea Astronomers inspire 7000+ students, teachers & families on the Big Island.

Thank You Again Gemini Observatory, State of Hawaii Department & Board of Education, Janice Harvey, Peter Michaud, Andolie Marten, Valerie Takata, Sponsoring Observatories & Local Businesses, Author & Gemini Observatory Astro-Photographer Extraordinaire Manuel Paredes, Astrophysicist Dr. Jeff Goldstein (for your passion & inspiration in creating Journey Week), and the entire Journey Education Team! See https://youtu.be/iC5JOaDGEg4
What a <u>privilege</u> to participate in one of the most beautiful places on Earth!

AND
Thank You! for sharing this passion for Astronomy & Science Discovery with me!

Pluto: My Journey from 9th Planet to 1st Kuiper Belt Object

Lowell Observatory's 13-inch Telescope with which Clyde Tombaugh Discovered former Planet Pluto. The Red Boxing Glove was added to prevent hitting one's head on the counterweight. *(Credit: the Author)*

Above: The Pluto Discovery Dome on the grounds of the Lowell Observatory in Flagstaff, AZ which houses the Discovery Telescope at left on the 2nd level. This is the tiny building in which Clyde Tombaugh discovered Pluto and transformed himself into a
World-Class Astronomer.
(Photo Credit: the Author)

Pluto: My Journey from Last to First

Discover Pluto's impact on the lives of those the little world has touched:
- **Percival Lowell – Flagstaff Observatory & Passion for All Things Pluto**
- **Clyde Tombaugh - Sketches which led to Pluto's Discovery**
- **Venetia Burney – Breakfast conversation which led to Pluto's Name**
- **James Christy – The "bump" which became Pluto's Largest of 5 Moons**
- **International Astronomical Union Astronomers – defining the word "planet"**
- **Alan Stern – 29-cent Challenge & Leading a Precision Team Fly-By in 2015**
- **Rock Stars and the World – participation in New Horizons' Ongoing Mission**

Much More than a Non-Fiction Photo History, Pluto's Fascinating True Story shares the Adventure Of Scientific Exploration and Discovery, and reveals practical Kuiper Belt Wisdom which the reader Can Leverage for a Successful Life on Earth!

www.ingramcontent.com/pod-product-compliance
Lightning Source LLC
Chambersburg PA
CBHW042024150426
43198CB00002B/55